T0188091

CAMBRIDGE COUNTY GEOGRAPHIES

General Editor: F. H. H. GUILLEMARD, M.A., M.D.

DERBYSHIRE

Cambridge County Geographies

DERBYSHIRE

by

H. H. ARNOLD-BEMROSE, Sc.D. ; F.G.S.

With Maps, Diagrams and Illustrations

Cambridge :
at the University Press
1910

CAMBRIDGE UNIVERSITY PRESS
Cambridge, New York, Melbourne, Madrid, Cape Town,
Singapore, São Paulo, Delhi, Mexico City

Cambridge University Press
The Edinburgh Building, Cambridge CB2 8RU, UK

Published in the United States of America by Cambridge University Press, New York

www.cambridge.org
Information on this title: www.cambridge.org/9781107637337

First published 1910
First paperback edition 2013

A catalogue record for this publication is available from the British Library

ISBN 978-1-107-63733-7 Paperback

NOTE

THE author is indebted to his wife Nellie Arnold-Bemrose for Chapters 20, 22, 23, 24, 25 and 28.

He wishes to acknowledge the information he has obtained from the *Victoria History of Derbyshire*, Dr Cox's *Churches of Derbyshire*, Mr Sandeman's paper on " The works of the Derwent Valley Water Board " and Kelly's Directory.

CONTENTS

ILLUSTRATIONS

ILLUSTRATIONS

MAPS

1. County and Shire. The Origin of Derbyshire.

In his preface to *The Making of England*, published thirty years ago, the historian J. R. Green truly remarks that "Archaeological researches on the sites of Villas and Towns, or along the line of road or dyke, often furnish us with evidence even more trustworthy than that of a written chronicle; while the ground itself, where we can read the information it affords, is, whether in the account of the Conquest or in that of the Settlement of Britain, the fullest and most certain of documents. Physical Geography has still its part to play in the written record of that human history to which it gives so much of its shape and form."

During the last thirty years much has been learned from physical geography in the deposits left in caverns and in the burying places of early Man. Men have struggled for mastery in our island against wild beasts, and have conquered some and domesticated others. Most of these wild animals have become extinct. Since then, men have fought with each other and our land has been invaded again and again by different races of people. In the struggle for existence the fittest have survived, and thus have been evolved the present inhabitants of England.

In treating of the geography of Derbyshire we may first pause to consider the difference of meaning of the words "County" and "Shire." Though these are modern terms they will help us to understand the relation which our present divisions of the kingdom bear to the ancient ones. The words "County" and "Shire," though now used as equivalent terms, have a very different origin and carry us back to different conditions in the political development of our country. The word "Shire," of Anglo-Saxon derivation, meaning a portion *shorn* or cut off from some larger division of land, implied as a rule a part of one of the great Anglo-Saxon kingdoms, though in some cases it was also applied to quite a small division of a district or even of a town, so that there were once, for example, six small "shires" in Cornwall, and there are still seven "shires" in the city of York. The word " County " is of Norman date and meant the land belonging to or ruled by a *Comte* or Count. More recently the word county acquired the same meaning as shire, so that now we speak indifferently of Derbyshire or the county of Derby. We cannot however call the county of Sussex, Sussexshire, because the historic counties and shires owe their origin to different causes. Each name is a survival of the ancient distribution of tribes which united to form the English people. Middlesex, Sussex, and Essex, for example, define the areas of three Saxon kingdoms, but Derbyshire is a share or portion of the kingdom of Mercia and is one of those counties which take its name from the county town.

We will now consider the origin of the word Derby which has thus given its name to our county.

For about 150 years after the first coming of the English, the Peakland and the northern parts of the present county of Derby were held by the Celts or Welsh. The few Mercians who settled in the hilly parts of Derbyshire were called Pecsaete, or Settlers in the Peak, so that the part of England which is now called Derbyshire narrowly escaped being called Pecsetshire after the fashion of Dorsetshire or Somersetshire. The word Derby we owe to the Danes. The Saxon name was Northweorthig, or Northworth, as it would be now written. The Danes, probably attracted by the Derbyshire lead-mines, overran the county and built a fort at Northweorthig, from which place the valley of the Derwent and its tributary valleys made access to the lead mining districts more easy. The Danes changed the name to Deoraby, which at a later date was abbreviated to Derby.

The name Derby is supposed by some writers to indicate a settlement by the uncleared deer-forest, but it is more probable that the name was derived from words expressing a settlement by the water (*dwr*). The first mention of Deorabisair, now Derbyshire, occurs only two years before the Norman Conquest.

2. General Characteristics. Position and Natural Conditions.

Derbyshire is an inland county near the centre of England bounded by Yorkshire, Cheshire, Staffordshire, Leicestershire, and Nottinghamshire. Its distance from the sea and the hilly nature of its surface made it for many centuries more or less inaccessible.

Though mountainous regions have now a fascination for lovers of nature, there was a time when people looked upon them with horror and only fled to them for refuge from their enemies. A writer about 150 years ago thus records the experience of some travellers on horseback on their arrival at Dovedale : " Proceeding towards the edge of the plain, they came to a precipice of an astounding height from which was a stupendous view into a deep valley, the hills rising on the opposite side covered with wood nearly half a mile perpendicularly."

Though the county with its rivers available for water power is adapted to many industries, few of them were fully developed owing, no doubt, to its inland position and the difficulty of communication. The various industries were with few exceptions only carried on to meet local needs. The earliest and most important of them was that of lead-mining, but, owing to the low price of the metal, this has of late years declined. With this exception Derbyshire was mainly an agricultural county. The introduction of canals, and improvements in the making

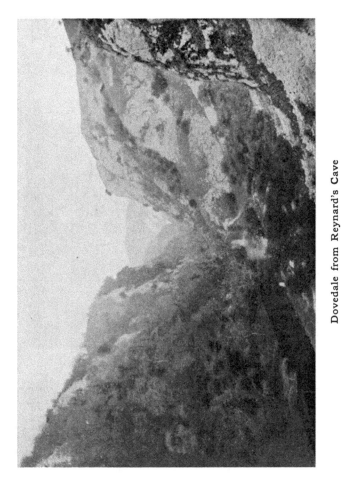

Dovedale from Reynard's Cave

and mending of roads, and the later communication by railways, cheapened the transit of goods and caused a rapid growth in the industries and population. The utilisation of coal and the manufacture of iron created quite a new and increasing industry in the county, which was rich in both these minerals.

Upperdale, near Monsal Dale Station on the Wye

During the nineteenth century, although the population as a whole increased, the number of persons engaged in cultivating the land became less, so that fewer persons were employed in agriculture in 1901 than in 1841. At present the railway and engineering works, coal-mines, quarries, and various factories, employ a large number of inhabitants, and Derbyshire, which at one time was an agricultural county, is now essentially a manufacturing one.

Derbyshire is noted for its beautiful and varied scenery. The uplands, or hilly northern portion of the county, form the southern spur of the Pennine Chain, the backbone of England, and rise to a height of more than

Monsal Dale

2000 feet above sea-level. The Peak District, the narrow dales and gorges in the limestone in Chee Dale, Dovedale, and at Matlock, and the various limestone caverns, are visited annually by a large number of people. The county is also rich in Prehistoric and Roman remains. Haddon Hall, Chatsworth House, Wingfield Manor, as well as numerous ecclesiastical buildings, are objects of interest. Lastly, warm mineral springs at Buxton and Matlock Bath have made these places for many years noted as health resorts, to which the surrounding fine scenery is an added attraction.

Chee Dale

3. Size. Shape. Boundaries.

Derbyshire has a rather irregular outline. Its greatest length from north to south is about 55 miles, measured from near Woodseats to Measham, and the greatest breadth from near New Mills in the west to the Nottingham border near Whitwell is 37 miles. The county encloses an area of 658,885 acres, or nearly 1030 square miles, and is about the same size as Cheshire.

About half the counties of England are larger than

Derbyshire and the remainder smaller, so that in point of size Derbyshire may be considered an average county.   Its shape is so irregular that it is difficult to describe it concisely.   The irregularity is mainly due to the fact that rivers form about three-fourths of the boundary.   The county is broad in the north and tapers towards the south.   While, as we have seen, its greatest

Dovedale

breadth in the north is 37 miles, the distance from the Dove near Tissington on the west to Pye Bridge on the Erewash in the east is only 18 miles, and in the southern part of the county, the breadth at the latitude of Burton-on-Trent is only six miles.

Derbyshire is bordered on the north by Yorkshire, on the east by Nottinghamshire, on the south-east by

Leicestershire, on the west by Staffordshire, and on the north-west by Cheshire. If we examine a map of the county, we soon notice that the following rivers form about three-quarters of the boundaries, viz. the Etheroe on the north-west, the Goyt, the Dove, and the Trent on the west, and the Erewash and the Trent again on the east. The remaining portions of the boundaries are mainly artificial.

A peculiarity connected with Derbyshire and the neighbouring counties may be referred to in this chapter. On old maps portions of Derbyshire appeared like islands in Leicestershire, and were entirely separated from Derbyshire geographically, though for administrative purposes they formed part of Derbyshire. A similar occurrence is found in other counties. Under the Local Government Board Act of 1888 it was provided that these outlying portions may be added to the district in which they are situated if both parties interested in the locality agree to the amalgamation. Portions of Leicestershire were accordingly transferred to Derbyshire, and portions of Derbyshire to Leicestershire in the year 1897. In the year 1894, portions of Nottinghamshire were transferred to Derbyshire, and the Derbyshire portion of Croxall, Stapenhill, and Winshill were transferred to Staffordshire. The boundary of the county had been previously altered by a part of New Mills being annexed from Cheshire and a part of Burton-on-Trent being transferred to Staffordshire.

Derbyshire   Goyt Valley near Buxton   Cheshire

4. Surface and General Features. Peakland and Lowland.

A clear idea of the chief features of the surface of Derbyshire will enable us better to understand the effect they have had upon the history and development of the inhabitants.

The surface of Derbyshire is very varied in character. An examination of the physical map at the beginning of this volume, coloured according to the heights of the land above sea-level, will show that the county may be divided naturally into two main portions—the uplands and the lowlands. These differ not only in altitude, but in many other ways.

The uplands, or the Peak District, as they are called, form the southern end of the long ridge called the Pennine Chain which extends through a large portion of England. The lowlands comprise the southern part of the county. The lowest parts in the south are slightly more than 100 feet above sea-level, whilst the highest part is Kinder Scout or the Peak, more than 2000 feet above the sea. The Peak is not, as its name implies, a point, but a nearly flat plateau or tableland, higher than any other portion of Derbyshire, which is situate to the north of Castleton and the Edale Valley. The name High Peak is applied to one of the hundreds or divisions of the county and the name Low Peak to the wapentake of Wirksworth.

The finest scenery is in the uplands. The Mountain Limestone with its outlines generally smooth and well rounded, with its deep narrow dales and gorges, presents a marked contrast to the wild moorlands and escarpments of the Millstone Grits by which it is surrounded. The latter attain a greater height than the limestone. The uplands are sparsely populated; the absence of hedges and the division of the fields by stone walls are very noticeable. In the lowlands the fields are flat, many-acred, and divided

Hope Valley, Castleton

by hedges, so that there is a great contrast between the two portions.

This difference of feature means a difference in the means of locomotion, and in the amount of rainfall and snow. In the uplands, as has been well said, the "land is mountainous, rocky, and wind-swept, and winter longs to linger far beyond its legitimate time."

There are extensive moorlands in Derbyshire, some of which once formed part of the ancient forests. The

King's Forest of the High Peak was a wild region in the thirteenth century. Its bounds began on the south at Goyt, went down the rivers Goyt and Etheroe, thence by Langley to the head of the river Derwent, to Mytham Bridge, Bradwell, Hucklow, and Tideswell; thence to the river Wye, and up the Wye to Buxton, so that the forest included the whole of the north-western portion of the county.

The Peak Forest, from which the village of Peak Forest derives its name, does not appear to have been well wooded at any time, but Duffield Forest, or Duffield Frith, in the southern part of the county, about five miles north of Derby, was more fertile and covered with woods.

5. Watershed and Rivers.

Some of the rivers in Derbyshire and on its borders flow west into the basin of the Mersey and others east into that of the Humber. The high land which forms the division between the two basins is called the watershed or divide, though sometimes this word watershed is used for the sloping ground down which the water flows on either side of the divide. In order to understand the position and nature of the river basins and drainage areas of a country we must know something of its physical conformation. The great watershed of Central England passes through the higher parts of Derbyshire, and the greater portion of the county lies to the east of this

divide. The actual water-parting passes by the Cat and
Fiddle (Axe Edge), near Buxton, along Rushup Edge
and Cowburn, a little to the east of Kinder Scout, the
western flank of the Peak ; thence it runs northward till
it meets the ridge formed by the northern outcrop of the
Millstone Grit, which it follows for about four miles, and

Ashop clough, Kinder Scout

then strikes away northwards beyond the boundaries of
the county. The streams rising on the west of this line
flow into the Goyt or Etheroe and find their way into
the Mersey. Those on the east flow into the Derwent,
the Don, or the Dove, and finally into the Humber.
The Derwent is the most important river in Derbyshire :
it has the largest drainage area (290,000 acres), and is

65 miles in length. It rises in the moorlands on the borders of Yorkshire and Derbyshire and forms the boundary between these counties for a short distance, but continues for the remainder of its course in the county. The Alport and Ashop join at Alport Bridge and enter the Derwent at Ashopton. The Noe flows along the wide dale of Edale and joins the Derwent at Mytham Bridge.

River Derwent, Hathersage

The Derwent then flows south, passes through Chatsworth Park, and is joined by the Wye near Rowsley Station. It then flows along the broad valley of Darley Dale and at Matlock enters the gorge in the Mountain Limestone, through which it flows as far as Willersley. The river then winds along more open dales to Derby, and thence takes a tortuous course through a wide alluvial valley to the Trent on the south-east of the county.

Other tributaries of the Derwent, in addition to those mentioned above, are the Amber, which rises on Darley Moor, the Ecclesbourn River and Bottle Brook coming from the Wirksworth valley, and the Markeaton Brook, which flows through Derby.

During its course through the limestone gorge at Matlock, the Derwent receives no tributaries, but the volume of water is very much increased by springs from the Mountain Limestone. In the Derwent basin there are many old lead-mines which were drained by levels or "soughs." Meerbrook Sough (now utilised for the water-supply of Heanor and Ripley) in 1868 yielded over 16 millions of gallons a day, which flowed into the Derwent.

The river Wye rises near Buxton on the northern slope of Axe Edge, and collects water in Buxton from a number of tributary streams both north and south. At Wye Head an underground stream issues from the limestone and flows through the gardens, where it is joined by the Serpentine, a stream which rises near Burbage. The Wye has been well described as "a singularly romantic river, running in deep rocky ravines, its clear stream sparkling along a confined and rugged bed." Its course is at first through the fine gorges of Chee Dale, Miller's Dale, and Monsal Dale, and then through a more open country at Bakewell. During the remainder of its course the river winds about considerably in the neighbourhood of Haddon Hall.

The Dove, an important tributary of the Trent, rises on the eastern slope of Axe Edge and drains 95,000 acres.

High Tor, Matlock Bath: River Derwent

It is one of our most beautiful streams, and passes through very fine scenery. It runs through the narrow gorge of Dovedale and then emerges into a broad and fertile valley, flows through Rocester and Tutbury, and enters the Trent at Newton Solney. Throughout the greater part of its course of 45 miles—during which it falls over

River Wye, Water-cum-Joly, near Miller's Dale

1500 feet—it forms the boundary between Staffordshire and Derbyshire. It has for many years been noted for its fishing.

The Erewash, which has a course of about 20 miles, forms part of the boundary between Derbyshire and Nottinghamshire, and flows into the Derwent.

The Rother, which rises in the N.E. part of Derby-

Ilam Rock, Dovedale

shire, flows through Chesterfield into the Don, and finally becomes part of the Ouse. The Rother drains in Derbyshire an area of over 88,000 acres.

The Trent flows through the county for a short distance and also forms part of the boundary of Derbyshire, on the west separating the county from Staffordshire for some eight or ten miles, flowing through Burton, and on the south-east dividing it from Notts for about half this distance.

The Goyt and Etheroe separate Cheshire from Derbyshire, drain about 54,000 acres in Derbyshire, and flow into the Mersey. The Goyt rises on Axe Edge, and flows down a picturesque ravine during the earlier part of its course.

6. Derwent Valley Water Scheme.

The account just given of the rivers of Derbyshire would not be complete without mention of the Derwent Valley water scheme. Under this joint scheme the upper parts of the area drained by the Derwent have been bought for the purpose of supplying water to Sheffield, Derby, Nottingham, and Leicester, and to the County of Derby. The Derwent and the Ashop rise in this area, which contains 31,946 acres, or 50 square miles of land, and lies at a height of between 500 and 2000 feet above sea level. The rainfall varies from 38 inches in the southern part to 60 inches in the more elevated regions, and the average for the whole is about 49 inches.

The Derwent Valley Water Board was established in

1899 in order to reconcile the claims of the authorities of the four large towns and the county above mentioned.

The method of obtaining the water is by making five large reservoirs, three on the river Derwent—called the Howden, Derwent, and Bamford reservoirs—and two on the river Ashop, the Haglee and Ashopton reservoirs. It is estimated that the Derwent reservoirs will have an area of 594 acres, and contain over six thousand million gallons of water, whilst those in the Ashop valley will have an area of 319 acres, and hold more than three and a half thousand million gallons of water. The ultimate scheme, in addition to the reservoirs, includes the making of about 100 miles of aqueduct for distributing the water, about 20 acres of filter-beds at Bamford, and a service reservoir at Ambergate.

On account of its magnitude the work has been divided into three parts or instalments, which will be carried out as they are required by the people for whose benefit they are intended. The first portion, consisting of the Howden and Derwent reservoirs, was commenced eight years ago and will probably be complete in 1912. It is estimated that it will deliver 12 or 13 million gallons of water a day and will cost $2\frac{1}{2}$ million sterling.

The reservoirs are formed by building large dams of sandstone and concrete across the valley. An idea of the magnitude of the work and of the method of carrying it out may be gathered from the following particulars and from the accompanying illustration of the Howden Dam, photographed in 1909.

The largest reservoir, called the Bamford reservoir,

will require a dam 1950 feet in length and 95 feet in height. The Howden and Derwent dams, which are nearly complete, are 1080 and 1110 feet in length, and 117 and 114 feet in height respectively, and will together store 3886 million gallons of water.

The whole of the work of constructing the dams is administered by the Derwent Valley Water Board instead of being let to contractors. The first acts of the Board were to build a railway seven miles in length from the Midland Railway at Bamford to the site of the Howden reservoir, and to construct the village of Birchinlee which had a population of 869 in December, 1908.

The stone used in making the dams is millstone grit, obtained from a quarry at Grindleford on the Midland Railway; and it is estimated that 1,200,000 tons will be required for the two first dams. The position for reservoirs of this enormous size was determined by the drainage area for the supply of water, and the nature of the rocks was necessarily a secondary consideration.

The foundations of the Howden and Derwent dams are in the shales and sandstones. These beds of rock are much contorted and crushed, and the ground is not watertight. It was, therefore, necessary to make what is called a watertight "curtain," i.e. a six foot width of concrete beneath the dam for its whole length and into the hills at each side. This curtain in the Derwent dam extends 55 feet down below the base of the dam; so that, although the dam is only 114 feet high, the distance from the foundations of the curtain wall to the top of the dam (including the dam foundations, 54 feet) is 212 feet.

Howden Masonry Dam

The main aqueduct extends from Howden to Ambergate reservoir and is 30 miles in length. The first instalment of the Ambergate reservoir, which is for regulating the supply to the towns, will hold 30 million gallons of water. The water will be distributed from this reservoir to Leicester, Derby, Nottingham, and the County of Derby, by means of pipes. The whole of the works, including the three instalments, are expected to cost £6,000,000, and the additional cost of carrying the water to the four towns will be slightly over one million sterling.

7. Geology. (*i*) Sedimentary Rocks.

In Chapter 4 we saw how the northern part of Derbyshire differed from the southern and eastern parts. Not only are there differences of elevation, however, but also those of soil and of the occupations of the people. In some places the hard rock comes to the surface and is quarried as limestone or sandstone, in others coal is being raised from below the surface of the earth; in others again clay forms the surface of the land, and is dug out and used for making bricks and tiles. There is a reason why one part is mountainous, another level, one thinly peopled and agricultural, another populous and industrial, one barren and another fertile. These variations are due to the differences between the rocks of which the ground is composed, and if we would learn why the rocks affect the shape and nature of the ground we must turn to geology for the explanation.

The word rock is used for any natural stone, whether it be hard or soft. Thus limestone, sandstone, clay, mud, and coal are all called rocks by the geologist. The greater portion of the rocks in Derbyshire are what are called sedimentary, and have been laid down in water just as gravel, sand, and mud are deposited by our rivers and lakes, and as limestone mud is laid down at the bottom of the sea.

The sedimentary rocks of Derbyshire consist mainly of clay, shale, sandstone, limestone, and gravel.

A visit to one of the numerous quarries in Derbyshire or a walk up one of our limestone dales will teach us that the rocks are arranged in beds or layers of varying thickness, one above another. The materials vary from bed to bed so that, as at Cromford station, limestone alternates with shale. It will easily be understood that the beds at the bottom of the quarry are the oldest, being necessarily deposited before those above them. This arrangement of one bed or stratum above another tells us the order in which the beds have been laid, the oldest being at the bottom, and the newest at the top.

The oldest beds in Derbyshire consist of Mountain Limestone, these are followed by the Limestone Shales, Millstone Grits and Coal Measures, and together form what is called the Carboniferous or coal series. Above this series we have the Permian or magnesian limestones and sandstones, the Bunter Conglomerates and Keuper Marls or clays with gypsum; later, the Pleistocene cavern deposits, glacial drifts, and clays; and later still the alluvium, peat bogs, calcareous tufa, and stalactitic formations.

GENERAL LIST OF THE ROCKS OF DERBYSHIRE.

QUATERNARY	Recent		Superficial deposits of Historic Iron, Bronze, and Neolithic Ages
	Pleistocene		Cavern deposits, Glacial Drift, Sands and Clays of Palaeolithic Age
TERTIARY	Pliocene		Cavern deposit of Doveholes
SECONDARY	Triassic	Keuper	Red Marl with Gypsum Waterstones
		Bunter	Pebble Beds or Conglomerate Lower Mottled Sandstone
PRIMARY	Permian	Magnesian Limestone Series	Marls and Sandstones Lower Magnesian Limestone Marls and Sandstones
	Carboniferous	Coal Measures	Middle Coal Measures Lower or Gannister Series
		Millstone Grit	Beds of Grit divided by Shales
		Limestone Shales	Shales with thin beds and nodules of Limestone
		Mountain Limestone	Limestone with Chert, thin Shales and Clay partings, and contemporaneous and intrusive Igneous Rocks

The rocks earlier than the Carboniferous are not seen because we have not reached the bottom of the limestone in Derbyshire. The Jurassic, Cretaceous, Eocene, and Pliocene rocks are absent from the county, though deposits of Pliocene age have been found in a cavern at Dove Holes.

The table overleaf gives a list of the strata found in Derbyshire. The thickness of some of the beds is unknown, whilst that of others varies in different parts of the county. The table denotes therefore only the relative ages or order of succession of the deposits.

The sedimentary rocks were deposited in more or less horizontal layers, and some of the beds—as those at Chee Tor near Miller's Dale—remain horizontal, but in various parts of the county, in quarries or natural sections, we can see that the beds of rock are inclined and in some cases have been bent into arches and troughs. The amount of inclination is called the dip.

At Cromford station the beds are not level or horizontal, but dip towards Derby and under the Black Rocks. At Matlock Bath opposite the High Tor the limestone beds dip rapidly, or at an high angle towards Derby, but as we walk to Matlock Bridge they become horizontal, as seen in the face of the High Tor, then roll over and dip towards Matlock Bridge to the north. What we see is the section of a dome or inverted basin of beds of rock, which is called an anticline. If instead of going to the Bridge we proceed by road to Cromford, we see that the limestone beds first dip towards Matlock Bath station, and then rise along the Lovers' Walk, roll over,

Chee Tor

(showing thick horizontal beds in Mountain Limestone)

and dip in the direction of Cromford, forming another
anticline.   Between the two anticlines, where the limestone beds dip into the ground, we have what is called
a syncline.   We thus learn that the rocks, owing to
lateral or side pressure, have been crumpled in some parts

Syncline and Anticline in Shales and Limestones,
L. and N. W. R. cutting, Tissington, near Ashbourne

of the county.   The figure given above shows a well-
marked anticline, and a syncline to the left of it, in the
railway cutting at Tissington.

The structure of the northern part of the county is
shown in the figure on the opposite page which is a rough
section across the Pennine Chain.   The beds are folded into

N.W.　Buxton　Bakewell　Ashover　Stretton　S.E.

Sketch-section across Derbyshire

(a) *Permian Limestone.*　(b) *Permian Sandstone.*　(c) *Coal Measures.*　(d) *Millstone Grit.*
(e) *Limestone Shales.*　(f) *Mountain Limestone*

a broad irregular dome, so that the lowest beds are in the middle of the area, dip east and west, and are covered by beds higher in the series. To the east and west of this dome are the synclinal troughs which form the coal-fields.

There are several places in the county in which the Mountain Limestone comes to the surface. But the largest mass of limestone forms an irregularly-shaped inlier measuring 20 miles from north to south, and 10 miles from east to west. Roughly, the beds dip away from the mass in every direction, the rocks on the east dipping gently beneath the shales. On the west the dip is greater, and the rocks are thrown into numerous folds, and often broken. The thickness of the limestone in Derbyshire is not known, the beds at the base not having been reached. The lowest beds seen are near Pig Tor tunnel in the valley of the Wye, and are about 1800 feet down in the limestone series.

The limestone varies in structure, composition, and colour. It is often an almost pure carbonate of lime. The upper beds are generally thin, and contain bands and nodules of chert. The limestone is distinguished by the number of fossil contents, which are mineralised remains of organisms living in the sea at the time the limestone was deposited.

The limestones are succeeded by a series of shales, with thin limestones and limestone nodules, termed the limestone shales. These are followed by shales and sand-stones.

Above these we have the Millstone Grit series, which

has been divided into five divisions by the Geological Survey.   They are found in the northern part of Derbyshire, and on the east and the west side of the Pennine Chain.   They extend as far south as Little Eaton.   The outcrop of each sandstone bed forms a long ridge with a sloping surface on one side in the direction of the dip,

Black Rocks, Cromford (*Millstone Grit*)

and on the other side a steep face or escarpment which is nearly vertical.   These escarpments are locally known as "edges," and form well-marked features in the landscape. Amongst the finest of them are Curbar, Froggatt, Bamford, and Derwent Edges on the eastern side of the county, and the "Black Rocks" near Cromford.

The Coal Measures lie on the east and west of the

Pennine Chain. In Derbyshire they are divided into the Middle Coal Measures about 2300 feet in thickness, and the lower or Gannister series, which is about 1000 feet. The Middle Coal Measures consist of sandstones, shales, and clays with ironstones and coal seams. The Gannister series is made up of flagstones and shales, with thin coal-seams, under which are floors of gannister. The seams of coal vary from 2 to 7 feet in thickness.

The fossils in the Coal Measures indicate a great profusion of vegetable growth during the time when they were formed. The flora, consisting of some hundreds of forms, has only distant representatives to-day in the tree-ferns of tropical swamps. The seams of coal are composed of compressed and mineralised remains of this vegetation.

In the east of Derbyshire, the Permian rocks, consisting of limestones, sandstones, and marls, have been deposited upon the upturned edges of the Coal Measures, and are found in a narrow strip running north and south. They were probably formed in isolated basins or inland salt lakes. The Permian rocks of Derbyshire consist mainly of the lower magnesian limestones and sandstones. The scenery of this limestone somewhat resembles that of the Mountain Limestone, but is on a much smaller scale.

The Triassic rocks have been divided into the Bunter and the Keuper. The Bunter in Derbyshire consists of pebble beds or conglomerate, and the lower mottled sandstone. It is found in several isolated patches. The largest extends from Ashbourne by Mugginton to Quarndon, near Derby. It is found at Norbury and Brailsford, and further south near Breadsall and Morley Dale, and at Sandiacre in the Erewash valley.

The Keuper beds (or new Red Marl) overlie the Bunter. They occupy a large tract of country south of Ashbourne, Breadsall, and Sandiacre, and stretch across the county in a direction from east to west. The Upper Keuper consists of red marl and shale with micaceous sandstone (called skerry) and irregular bands of gypsum. The Lower Keuper consists mainly of sandstones.

The Keuper beds were deposited on the more or less tilted edges of the Carboniferous rocks. They cover in one place Millstone Grit, in another the Yoredale rocks, and in another the Mountain Limestone. Hence, before the Keuper period, earth movements took place which raised the older rocks and exposed them to the action of the weather.

The Pliocene and Pleistocene deposits of Derbyshire owe their preservation to the fact that they have been washed into caverns and thus protected from the denudation or wear and tear of the rocks above them. A brief description of these deposits will be found in Chapter 10.

The later Pleistocene or glacial deposits consist of clays, sands, and gravels. Boulders varying in size and character are often found embedded in the Boulder Clay. Some of them consist of rocks derived from the district, others are foreign to it, and must have travelled hundreds of miles from the places where they once formed part of the natural rock. These boulders are frequently much scratched, grooved, and polished from having been

Boulder Clay, Crich

pressed and rubbed against the rocks of the country over which they passed, and when the rocky floor is laid bare by the removal of the clay it has been found to be covered with scratches and grooves whose bearings indicate the direction from which the boulders and clay have been brought. A few years ago, scratches in a N.N.W. direction were seen on the limestone floor below the Boulder Clay at Crich.

8. Geology. (*ii*) Igneous Rocks.

There are some rocks in Derbyshire, locally called "toadstone," which have had a very different origin from that of the sedimentary rocks described in the last chapter. Several good exposures of these rocks may be seen in travelling by rail from Derby to Miller's Dale. Immediately after leaving Matlock Bath station and just before entering the High Tor tunnel, a bed of dark-coloured rock is seen on both sides of the railway cutting. It is about 70 feet in thickness, with beds of limestone above and below it. Just before we arrive at Miller's Dale station there is a bed of similar dark rock at the bottom of the dale on our right with a bed of white limestone above it; whilst in the wall of the cutting on our left hand we can see traces of a bed of similar dark rock, between which and that at the bottom of the dale is a stratum 150 feet thick of limestone beds. If we alight at Miller's Dale and walk a short distance up Priestcliffe Lane towards Taddington we soon enter the upper bed of dark rock. These beds have been traced through various parts of the county and sometimes extend over a number of square miles and are always interbedded with the limestone. That they were once lava streams is evident on examination of the rock, which is crystalline in structure,

Lava and Tufaceous Limestone resting on Limestone, Buxton Lime Co.'s Quay, near Miller's Dale Station

t = lava, a = tufaceous limestone, l = limestone

studded with numerous holes or vesicles, some of which have been filled with carbonate of lime and are called amygdaloids. There is another variety, also interbedded with the limestone, which is not a massive rock but made up of fragments, which vary in size and are arranged in thin layers. The fragments, when carefully examined, are found to be irregular in shape and composed of a volcanic glass with numerous small steam holes. They are due to the breaking up of molten rock in a volcano and are called lapilli, while the beds of rock are known as volcanic tuff[1].

Beds of tuff may be seen amongst other places at Litton near Tideswell, in Cressbrook Dale, Tearsall Farm, near Matlock, and in the Tissington railway cuttings. These lava streams and tuff beds tell us that whilst the limestone was being deposited upon the bed of the ocean, submarine volcanoes burst forth, welled out their lava streams, and deposited their tuff on the sea floor.

The volcanoes from which these outbursts came must have been buried under further accumulations of materials on the sea floor, but in a county like Derbyshire where the limestone is entrenched by deep valleys we might expect to find traces of the pipes through which the volcanic material came to the surface.

At Grange Mill, about five miles from Matlock Bath, are two dome-shaped hills with grassy slopes rising from the valley to a height of 100 and 200 feet respectively. They consist mostly of a grey rock with numerous green lapilli, whilst some parts are of coarser material.

The position of these hills and their relations to the limestones surrounding them show that they form the necks or stumps of old volcanoes composed of the material

[1] The term Volcanic Ash is sometimes applied to them but though in popular use is incorrect because the rocks have not been subjected to fire.

which has filled the pipe up which the rock came from the interior of the earth.

In the limestones which are seen on both sides of the valley is a bed of volcanic tuff about 90 feet in thickness. It is higher up in the series of limestone and was probably thrown out of these vents.

A good section of the tuff may be seen at Shothouse

Grange Mill Vents from the N.W.

Spring on the road from Grange Mill to Winster. Perhaps the largest vent in Derbyshire is Calton Hill near Taddington, which is composed of a dark green or black crystalline rock known as basalt.

There are other vents near Castleton, at Hopton, and at Kniveton. We also have in Derbyshire some igneous rocks of a later period than the Carboniferous volcanoes.

They may be popularly described as volcanoes which failed to reach the surface of the ground.

After the limestone had become consolidated into hard rock the molten material from the inside of the earth pushed its way up into the Mountain Limestone, across the beds and in some places between them, baking the limestone with which they came into contact and making it into a crystalline limestone or marble.

By the agency of denudation the rocks above them have been removed and valleys have been cut into them by rivers, so that we are able to find sections which show their relation to the beds in contact with them. These intrusive masses, as they are called, are composed of a coarse-grained crystalline rock, very hard and black or dark green in colour. In Tideswell Dale a sheet of this intrusive rock about 70 feet in thickness has baked the limestone below it into marble for a depth of some feet. At Peak Forest a similar rock has baked the limestone beds above it.

In Tideswell Dale and on Masson Hill near Matlock and at Water Swallows the intrusive rock is quarried for road-metal.

9. Caverns and Underground Drainage.

There are numerous caverns in Derbyshire in the upper beds of the Mountain Limestone which are interesting either because of the manner in which they have been formed, or from their connection with the underground drainage of the district, or lastly from the mammalian deposits which some of them contain. In some parts of the limestone district there are numerous vertical cavities or "swallow-holes" in the ground, which have been formed by water dissolving the limestone and carrying it away in solution. The hamlet of Water Swallows near Buxton no doubt owes its name to the number of

Entrance to Peak Cavern, Castleton

swallow-holes in its neighbourhood. The water disappears down these holes and the surface drainage passes underground and dissolves out of the solid rock a system of chambers and tunnels. The Peak Cavern at Castleton, which has a magnificent entrance, is an example of a natural cavern connected with a system of underground drainage.

The water which runs down Rushup Edge north of the road from Chapel-en-le-Frith to Castleton, instead of flowing down the valley in a westerly direction towards Chapel-en-le-Frith, enters the limestone along a line of swallow-holes near Perryfoot, at the boundary of the Mountain Limestone and shales. The water is finally discharged partly through the caverns, but largely by a

Middleton Dale

spring called Russett Well, and the combined stream, known as Peak's Hole Water, flows down the valley, joining the river Noe near Hope.

Another system of underground drainage occurs near Eyam. The water enters the limestone by swallow-holes and finds its way to the valley of the Derwent by way of Middleton Dale. The disappearance of the water down swallows often results in a dry valley which represents the old watercourse. Linen Dale near Eyam is one of these dry valleys, and Great Rocks Dale, through which the Midland Railway passes between Doveholes and Buxton Junction, is another. The Winnats and Cave Dale near Castleton are fine examples of dry valleys.

The Speedwell Cavern and the Blue John mine near Castleton are partly natural and partly artificial. The entrance to the Speedwell is at the foot of the Winnats. A level was driven into the hill to reach some of the lead veins and entered the New Rake vein at a distance of 750 yards from the entrance. The level now contains water and visitors are taken in a boat to the large narrow cavern, which extends to a great height and was hollowed out by underground waters. A solid platform has been built on the sloping floor of the cavern and the excess of water falls over into the lower part known as the "Bottomless Pit." This pit was explored by the Kyndwr Club in 1901, and the water at the bottom was found to be 20 feet deep and 63 feet below the platform.

The Blue John mine consists of large underground cavities connected by artificial passages, and derives its name from the variety of fluor spar known as " Blue John " which is obtained from it. The total distance of the winding passages is said to amount to more than three miles. Eldon Hole, a chasm 180 feet deep in the side of Eldon Hill near Peak Forest, which the writer and others explored in 1900, is about 100 feet long and 20 feet wide at the top, and at the bottom measures 36 feet by 29 feet. The floor when reached from the surface is found to be composed of loose angular blocks of lime-

The Winnats, Castleton

stone and from it a low archway opens out into a large cavern, the lowest part of which is 256 feet from the surface of the ground.

Amongst other caverns shown to visitors are the Bagshawe Cavern at Bradwell near Castleton, the High Tor, Cumberland, and Jacobs Caverns, Matlock Bath, and Poole's Hole, Buxton. As a rule the parts accessible to

Dove Holes, Dovedale

the public form but a small proportion of the whole, the passages sometimes extending to several miles in length.

Water charged with carbonic acid has the property of dissolving limestone and in this way the caverns have been formed. As the water evaporates the carbonate of lime which has been dissolved is re-deposited as beds of tufa, and in a cavern the drippings from the roof form stalactites hanging from the roof and stalagmites built up from the floor which often produce most beautiful effects. Large deposits of this rock have been formed by the warm springs at Matlock Bath. Here and in Via Gellia it is quarried for ornamental rock work.

The "Petrifying Springs" at Matlock Bath, which issue from the limestone, form deposits of carbonate of lime on any small objects placed in them and at the present day along their short course into the Derwent are forming tufa.

10. Caverns and their Mammalian Contents.

Many of the caverns of Derbyshire are interesting because of the records they contain of animals which existed in England, not only during the later Prehistoric period, but also during the older Pliocene and Pleistocene periods. These records consist of bones buried in clay or cave earth. In some cases a more ancient deposit has been carried by water to a lower level of the cavern, and subsequently the upper part of the cavern has been carried away by denudation or the action of the weather. In other cases, the animals have lived in the caverns or been taken into it by hyaenas. Still later, man and domesticated animals have lived in the cave, and their remains have been covered up by the deposits of clay and stalagmite. In this chapter we shall briefly consider some of the more important of the Pliocene and Pleistocene deposits.

The Dream Cave at Wirksworth was explored by Dean Buckland in the early part of last century. He found an almost perfect skeleton of a rhinoceros. The animal had fallen down an open swallow-hole, and was buried in the clay and loam introduced by a stream of water.

At Windy Knoll, near Castleton, is a swallow-hole in the limestone. It contained bones of the bison, bear, fox, wolf, and reindeer. Professor Boyd Dawkins considered that these remains point out one of the routes by which the bisons and reindeer passed from the east to the west of England, or from the valley of the Derwent into the plains of Lancashire and Cheshire.

On the north-eastern boundary of Derbyshire, four of the Creswell Caverns in the Magnesian or Permian limestone were explored by the Rev. J. M. Mello and Professor Boyd Dawkins. They contained remains of a Romano-British population, with bones of recent animals. Below these deposits were bones of the cave bear, hyaena, wolf, bison, deer, lion, mammoth, woolly rhinoceros, and the modern horse (*Equus caballus*), together with the tools of Palaeolithic man.

In 1902 a cavern in the limestone quarry at Hoe Grange, near Longcliffe, was broken into. The cavern was filled with clay and sand, which contained numerous bones. The writer obtained over 8000 specimens, of which 4545 were named by Mr E. T. Newton. They comprised some twenty-seven species of vertebrate animals. There was no evidence of the presence of man. The reindeer, which was present at Creswell, was absent from

Longcliffe, and its absence is the more remarkable in that it occurs in nearly all the lists of animals from Derbyshire caves. The rhinoceros at Longcliffe is different from the woolly rhinoceros which has been previously found in Derbyshire. The horse, though present in most of the Derbyshire caves, was absent from Longcliffe. The mammoth (*Elephas primigenius*) was found at Creswell, but was absent from Longcliffe, whilst *Elephas antiquus* was represented at Longcliffe by a milk tooth only.

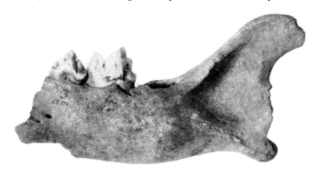

Left lower jaw of *Felis leo* showing milk teeth, about ⅔
(*From Hoe Grange Cavern, Longcliffe: Pleistocene*)

The fallow deer is supposed to have been introduced into Great Britain by the Romans, but the large number of bones of this animal found in the Longcliffe cavern mingled indiscriminately with those of the other Pleistocene animals proves that it existed in Britain in Pleistocene times. The occurrence of the lower jaw of a lion's whelp was said by Professor Dawkins to be "the most important recorded from any cave in this country."

A cavern at Dove Holes, 2½ miles north of Buxton, was broken into by the quarrymen prior to the year 1903, and the bones, which were described by Professor Dawkins, were those of the hyaena, the sabre-toothed tiger (*Machairodus*), the mastodon, *Elephas meridionalis*, *Rhinoceros etruscus*, *Equus stenonis* (probably the ancestor of the horse of Pleistocene age and the present day called *Equus caballus*) and *Cervus etueriarum*. These mammalia belong to the fauna of the Pliocene strata of Britain and the Continent.

Upper Canine of Sabre-toothed Tiger, about ½
(*From Dove Holes Cave: Pliocene*)

The Dove Holes cave "is the only Pliocene cave yet discovered in Europe, and is the only evidence yet available of the existence of the upper Pliocene bone caves which from the nature of the case must have been as abundant in Europe as those of the succeeding Pleistocene age." Professor Dawkins considers that the fragmentary remains in the cave at Dove Holes were derived from a den of hyaenas belonging to the Pliocene age, and that they were conveyed from a higher level into it by water, and that the cave at Dove Holes escaped the destruction by denudation because it was a sufficient distance below the surface of the ground.

Remains of Pleistocene mammalia, frequent in the river gravels of the southern counties, have been found in at least one place in Derbyshire. In 1896, the author and Mr R. M. Deeley obtained the greater portion of the skeleton of a hippopotamus, together with part of the breast-bone of an elephant and of the femur of a rhinoceros, in the Derwent gravel at Allenton, immediately to the south of Derby. These bones are now in the Museum at Derby.

11. Natural History.

There are several lines of evidence which lead to the conclusion that what we call the British Islands formed part of the continent of Europe at a late geological period.

The bones of extinct animals found in the caverns of Derbyshire and other parts of England, and in the old river gravels, are similar to those found on the continent and on the bed of the North Sea, especially on the Dogger Bank. When these animals invaded Great Britain, the North Sea could not have existed and the English Channel also, instead of sea, must have been a broad plain or river valley. The land must therefore have been some 300 feet above its present level. The raised beaches on some parts of our coasts also point to an elevation of

the land, while on the other hand the remains of forests
now sunk beneath the sea and only to be seen at low
water show that a sinking of the land has since taken
place. The fact that the number of species is greater
on the Continent than in Britain, and greater in Britain
than in Ireland, shows that Britain was severed by sea
from the Continent before all the European species had
time to establish themselves with us, while even before
this Ireland must have been separated from Great Britain.

The distribution of the flora of Derbyshire is mainly
determined by the climate and the soil. The former
varies to a great extent with the altitude, and the latter
depends upon the rocks from which the soil has been
produced. Different plants thrive best on different soils
and as we proceed from lower to higher ground, the less
hardy plants die out. There is therefore a very rich and
varied flora in a county like Derbyshire. The number of
species of flowering plants which have been noted in the
county is about 1000. The richest part is the Peak
district, including the Mountain Limestone and Millstone
Grit, in which are found subalpine and bog plants. In
the dales are many plants peculiar to the Mountain Limestone. The fragrant lily-of-the-valley flourishes in the
loose dry gravel on the steep limestone slopes of the Via
Gellia and Monsal Dale, the yellow heartsease and white
saxifrage grow well in the banks and fields on the
Mountain Limestone, and the small yews and juniper
bushes form a marked contrast to the white limestone
crags on which they grow in many of the dales and
gorges. The central area, including the Coal Measures,

is the poorest in plants, and the southern or lowland part is rich in such plants as are found in similar districts of England.

In the parks of Derbyshire are many fine oaks and Spanish chestnuts, as well as beech, sycamore, elm, horse-chestnut, and lime. In Darley Dale churchyard is an ancient yew tree 32 feet in circumference, dating from Saxon times.

The wild animals of Derbyshire of the present day differ little from those of other counties. The weasel is common in all parts of the county. The stoat is rare in the northern parts, and the fox survives in considerable numbers in the southern parts, which is hunted, but is killed in the moorlands, where it is exceedingly rare. The badger is becoming more rare and the marten and polecat are extinct. The otter exists in large numbers on the river Dove, especially in the lower portion, where protection is given to it by the landowners, but is scarce in the upper Dove and in the Derwent. The red deer, which is known to have existed in a wild state in the Forest of the Peak until about the year 1600, is now found only in the parks of Chatsworth, Hardwick, and Calke Abbey. There are about twelve herds of fallow deer in the county, and many of them may be descended from the wild fallow deer which inhabited the Peak and other forests. The herd at Stanton-in-the-Peak, near Rowsley, consists entirely of the black variety.

Derbyshire is not so rich in birds as some of the other counties, because its distance from the sea prevents the visits of maritime birds, and because the greater part of the county, with the exception of the Trent valley, is outside the main migration routes.

There is however in Derbyshire an overlapping of the northern and southern kinds of birds. The lowlands are a breeding place for such southern species as the nightingale, and the uplands for such birds as are rarely found breeding in the central plains of the Midlands. The ring ousel, a summer visitor to the uplands, and the meadow pipit, a resident, seldom breed below an altitude of 1000 feet, whilst the yellow wagtail and red-backed shrike nearly always breed at a height of less than 500 feet. On the bare uplands bird life is comparatively scarce.

12. Climate and Rainfall.

The climate of a country or district is, briefly, the average weather of that country or district, and it depends upon various factors, all mutually interacting, upon the latitude, the temperature, the direction and strength of the winds, the rainfall, the character of the soil, and the proximity of the district to the sea.

The differences in the climates of the world depend mainly upon latitude, but a scarcely less important factor is this proximity to the sea. Along any great climatic zone there will be found variations in proportion to this proximity, the extremes being "continental" climates in the centres of continents far from the oceans, and "insular" climates in small tracts surrounded by sea. Continental climates show great differences in seasonal temperatures, the winters tending to be unusually cold and the summers unusually warm, while the climate of insular tracts is characterised by equableness and also by greater dampness. Great Britain possesses, by reason of its position, a temperate insular climate, but its average annual temperature is much higher than could be expected from its latitude. The prevalent south-westerly·winds cause a drift of the surface-waters of the Atlantic towards our shores, and this warm-water current, which we know as the Gulf-stream, is the chief cause of the mildness of our winters.

Most of our weather comes to us from the Atlantic. It would be impossible here within the limits of a short chapter to discuss fully the causes which affect or control weather changes. It must suffice to say that the conditions are in the main either cyclonic or anticyclonic, which terms may be best explained, perhaps, by comparing the air currents to a stream of water. In a stream a chain of eddies may often be seen fringing the more steadily-moving central water. Regarding the general north-easterly moving air from the Atlantic as such a stream, a chain of eddies may be developed in a belt parallel with its general direction. This belt of eddies, or cyclones as they are termed, tends to shift its position, sometimes passing over our islands, sometimes to the north or south of them, and it is to this shifting that most of our weather changes are due. Cyclonic conditions are associated with a greater or less amount of atmospheric disturbance ; anticyclonic with calms.

The prevalent Atlantic winds largely affect our island

in another way, namely in its rainfall.   The air, heavily
laden with moisture from its passage over the ocean,
meets with elevated land-tracts directly it reaches our
shores—the moorland of Devon and Cornwall, the Welsh
mountains, or the fells of Cumberland and Westmorland
—and blowing up the rising land-surface, parts with this
moisture as rain.   To how great an extent this occurs is
best seen by reference to the accompanying map of the
annual rainfall of England, where it will at once be
noticed that the heaviest fall is in the west, and that it
decreases with remarkable regularity until the least fall
is reached on our eastern shores.   Thus in 1906, the
maximum rainfall for the year occurred at Glaslyn in the
Snowdon district, where 205 inches of rain fell; and the
lowest was at Boyton in Suffolk, with a record of just
under 20 inches.   These western highlands, therefore,
may not inaptly be compared to an umbrella, sheltering
the country further eastward from the rain.

The above causes, then, are those mainly concerned
in influencing the weather, but there are other and more
local factors which often affect greatly the climate of a
place, such, for example, as configuration, position, and
soil.   The shelter of a range of hills, a southern aspect,
a sandy soil, will thus produce conditions which may
differ greatly from those of a place—perhaps at no great
distance—situated on a wind-swept northern slope with
a cold clay soil.

It is interesting to know how rainfall is measured, and
how its distribution is recorded.   When a meteorologist
speaks of the mean rainfall at Buxton being 52 inches, he

means that if all the rain which falls upon a level piece of ground in Buxton during an average year could be collected without waste, at the end of the year it would form a layer of water which would cover the piece of ground to a depth of 52 inches. The weight of such a mass of water is very great, as may be realised from the fact that one inch of rain is equivalent to 100 tons of water on each acre. The rainfall of course varies from day to day and year to year. Thus in 1882, one of the wettest years on record, the rainfall at Buxton was 65·86 inches, whereas during 1887 it was only 32·38 inches.

The rainfall for a number of years in Great Britain has been measured and recorded by voluntary observers, whose records have been published annually in *British Rainfall* from 1860 to the present time. The results thus obtained are shown on the map illustrating this chapter. From this it is plain that the heaviest rainfall is in the Welsh and Cumberland hills and in Cornwall.

A similar map for Derbyshire was made by Dr Barwise, and published in 1899, in the "Report upon the Water-Supplies of Derbyshire" by himself and Mr J. S. Story. The mean rainfall varied from below 25 inches to over 50 inches. The heaviest rainfall in the county occurred near Woodseats in the Derwent basin and at Fairfield, near Buxton, whilst over the strip of land between Derby and the river Trent containing Willington and Alvaston the lowest rainfall was recorded.

In 1907 the highest rainfall in Derbyshire was at Bleaklow Stones, 2060 feet above sea level in the Peak

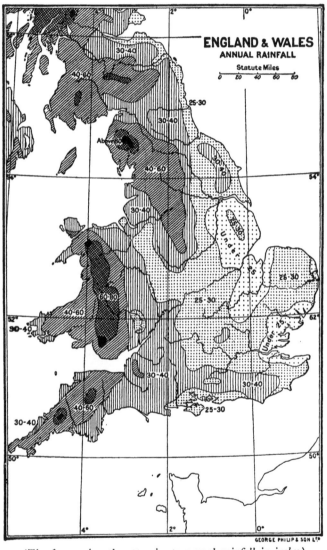

ENGLAND & WALES
ANNUAL RAINFALL
Statute Miles
0   20   40   60   80

30-40
40-60
25-30
30-40
Aberd...
30-40
40-60
30-40
Un... ...s
25-30
60-80
25-30.
40-60
52°
30-40
Under
30-40
30-40
40-60
30-40
30-40
Under
25-30

GEORGE PHILIP & SON Ltd

(The figures give the approximate annual rainfall in inches.)

district, where it measured 64 inches. At Derby it was 28·38 inches, or rather more than two inches above the average of 30 years.

In Derbyshire there are few records of temperature over a series of years compared with those of rainfall. The mean temperature for twenty years (1881–1900) at 9 a.m. at Buxton, 987 feet above sea level, was 46·1° Fahr., and at Belper, 344 feet above sea level, 46·8°. At both places July was the hottest month and January the coldest. The mean temperature at Buxton for 20 years (1881–1900) was 45·2° Fahr., and at Belper 47·3°.

Buxton is the only place in Derbyshire where regular records of sunshine have been made. The average of five years (1904–1908) was 1334 hours. The 1313 hours of sunshine at this place in the year 1907 may be contrasted with the 1666 hours at Cromer, the 1234 hours at Westminster, and the 894 hours at Manchester in the same year.

13. People and Population.

Little is known about the earliest inhabitants of Derbyshire. They were probably men of the Palaeolithic or Early Stone age, who lived by hunting and fishing at the time when England was joined to the Continent. They dwelt in caverns, and were without domestic animals or a knowledge of agriculture. These men were succeeded by those of the Neolithic or New Stone age who lived in rude huts in spaces which they had cleared in the forests. They had domesticated the goat, ox, sheep, dog, horse, and hog, had cultivated wheat and flax, possessed some knowledge of spinning and weaving, and understood the art of making pottery by hand.

When Julius Caesar invaded Great Britain he found that the Britons belonged to various races, used different languages, and were in different stages of civilisation. The people inhabiting Derbyshire were called by the Romans Coritani.

The Romans had three forts in Derbyshire, viz. Little Chester, Brough, and Melandra Castle. Derventio (Little Chester) was the most southerly of the system of the North British auxiliary forts held by the Romans. That this county was then of importance is shown by the Roman " pigs " of lead which have been found in Derbyshire, and the extensive system of Derbyshire roads in Roman times.

On the retirement of the Romans at the dawn of the fifth century, the Celts or Welsh were left in possession and retained a large part of the Peakland for 150 years. They were disturbed by hordes of Picts from Scotland, and by Saxons from Northumbria. The Saxons soon settled in parts of Derbyshire, and founded the kingdom of Mercia. Those who settled in the Peak were called Pecsaetas or Peak settlers, and the county might have been called Pecsetshire. Some of the Celts were driven away and others gradually became absorbed in the Saxon or Mercian population. Northweorthig (now Derby) was held by the Saxons for three centuries, and Repton shared with Tamworth the "Capitalship" of Mercia. Ethelfrith in the seventh century took a large part of the

Peakland from the Celts or Britons, and thus considerably extended his kingdom. The country was invaded by the Northmen or Danes, who were attracted by the lead, and then settled here. They changed the name Northweorthig to Derby, the termination "by" denoting its Scandinavian origin. They took the north part of the kingdom as one of their five "burghs," each ruled by its own earl.

There was much fighting between the Saxons and the Danes until in 941 A.D. King Edmund finally freed from Danish rule the five burghs and all Mercia "long time constrained by heathen men in captive chains."

The ancient or geographical county of Derby contains, according to revised returns, an area of 658,885 statute acres. Its population numbered 620,322 persons in 1901. In 1801 the population was only 161,567. It has therefore nearly quadrupled during the century. When the census was taken in 1901, there were 610,522 people in the Administrative County of Derby, 105,912 of whom were in the county borough of Derby. About three-fifths of the people live in towns or in urban districts and the remainder in villages or in rural districts.

The average number of persons to a square mile is 600 in Derbyshire, compared with 558 for the whole of England and Wales, so that the density of its population is slightly above the average. The census of 1901 shows that there were more males than females in Derbyshire. The former numbered 306,545 and the latter 303,977.

Of the 620,322 persons enumerated in the county of Derby, 431,803, or 69 6 per cent., were born within the county, 5310 were born in London, 3882 in Ireland, 2580 in Wales and Monmouthshire, 2321 in Scotland, and 997 in British colonies and dependencies. Persons of foreign birth numbered 1341 ; 859 of these were British and naturalised British subjects, and 482 foreigners. Only 211 persons were enumerated as having passed the night of the census of 1901 in barns, sheds, and caravans.

The main occupations of the people were as follows :— the men were chiefly engaged in the mining and metal industries, in agriculture, on railways, in house-building, and other trades ; while the women following occupations were mainly domestic servants, workers in cotton, hosiery, and lace, dressmakers and milliners.

There were 356 blind persons, 362 deaf and dumb, and 1692 lunatics and imbeciles in Derbyshire, 1202 of the latter being inmates of institutions. It should be noted that the numbers of persons enumerated as suffering from one or another infirmity are affected by the presence or absence of institutions in which many of such persons might be resident. Thus of the 198 persons enumerated as deaf and dumb in the county borough of Derby, 155 were in the Royal Institution for the Deaf and Dumb, the use of which is not confined to the inhabitants of the borough.

14. Agriculture—Main Cultivations, Woodlands, Stock.

The agricultural character of Derbyshire is as varied as its surface. The meadows on the banks of the Trent, Derwent, and Dove in the southern lowlands of the county provide a rich pasture, but the northern uplands are poor grazing land and in some parts yield only a scanty herbage. The red marl or clay and the gravels in the arable district south of Derby are productive, whilst the limestones in the north are generally unsuited for anything but permanent pasture. The coal districts in the eastern part of Derbyshire are mainly devoted to the getting of coal and to allied industries. The statistics of the Board of Agriculture and Fisheries for the year 1906 deal with 650,370 acres in the county of Derby; of which 35,274 acres are mountains and heath, and 3958 water.

The total acreage of cultivated land according to the returns made in 1906 was 489,322 acres, of which 450,128 acres are occupied by tenants and 39,194 acres are farmed by the owners. The total number of holdings is 11,481, of which 2392 are above one and less than five acres, 6129 are above five and less than 50 acres, whilst not more than 2870 are above 50 acres and only 90 above 300 acres. The average size of the Derbyshire farm is 42·6 acres. So that though Derbyshire is not a great agricultural county, the small farm system is not only common, but in many cases the farm has consisted of much the same acreage for several hundred years and has been farmed by the same family for many generations.

The vegetable products of Derbyshire—as well as of all the counties—are arranged by the Board of Agriculture under the following divisions :—Corn Crops, Green Crops, Sainfoin and Grasses for Hay, Other Crops, and Small Fruit. The portion of land which does not produce any of these crops is described as Bare Fallow, and in our county this comprised only 1749 acres in the year 1906.

The corn crops are grown on 41,206 acres and consist of wheat, barley, oats, rye, and beans and peas. Thus about one-fifteenth of the area of Derbyshire is devoted to these crops. Oats and wheat are the most important; barley being grown on 5034 acres only.

The green crops cover about 18,000 acres, or one thirty-sixth of the county. They consist of turnips and swedes, mangold, potatoes, cabbage, and vetches or tares.

A somewhat larger portion of the county, viz. 24,358 acres, or about one twenty-seventh of its area, is devoted to the growing of clover, sainfoin, and grasses. Nearly three-fourths of this produce is " for hay," the remainder is " not for hay," but the land is broken up in rotation. The largest portion of agricultural land in Derbyshire is used for Permanent Pasture, or grass not broken up in rotation. This area of permanent pasture measures no less than 402,857 acres, or nearly two-thirds of the entire area of the county. The proportion of arable land to permanent grass land is a little over one-fifth.

The woodlands of Derbyshire, though formerly more

extensive, covered only 25,852 acres in the year 1906. Coppice woods (or those which are cut over periodically and reproduce themselves naturally by stool shoots) and plantations form only a small proportion of the whole woods.

The animals reared in Derbyshire for various purposes are divided into four classes, viz. horses, cattle, sheep, and pigs. In the year 1906 cattle numbering 142,450 formed the largest class : the number of sheep was nearly the same as that of cattle, 140,773. The number of horses was 28,472, and of pigs 27,751. The greater number of horses are used for agricultural purposes, whilst the cows are reared to supply milk for the towns of Derbyshire and for Liverpool, Manchester, and Stockport.

In the south-western part of the county condensed milk is largely manufactured. In 1901, the Anglo-Swiss Condensed Milk Company started works which have had a great influence on the agriculture of this part of the country. The supply of milk is derived from Derbyshire and Staffordshire, and the maximum daily quantity is about 20,000 gallons in summer and 10,000 in winter. The annual output consists of about 15 million tins of condensed milk, which are mainly exported to the British colonies.

Cheese-making, which some fifty years ago was so great an industry, has become almost extinct except in large factories. The reason for this is that the railway system increased the facilities for selling milk in London, Liverpool, Manchester, and Stockport.

15. Industries and Manufactures.

Derbyshire in early times was mainly an agricultural county and, with the exception of lead-mining, industries were only worked to supply local needs. But since the seventeenth century, when the mining industries made a great advance, and in the nineteenth century, when railways came in and enabled coal to be applied more largely to industrial purposes, there has been a rapid advance in industry and a decline in agriculture.

The census returns taken every ten years show the changes which have taken place. The number of people engaged in agriculture in 1841 was over 18,000; this rose to 26,000 in 1861, and fell to below 16,000 in 1901. The numbers engaged in the textile trades rose from 11,000 in 1841 to 22,000 in 1861, and fell to below 13,000 in 1901, whilst those employed in the hosiery and lace industries fell from 8000 in 1841, to 4000 in 1871, and rose to nearly 9000 in 1901.

The mining and iron trades have during the same period rapidly increased. The iron trades employed about four times as many persons in 1901 as in 1841. The increase in mining was much greater. The number of persons employed in mining in 1901 was more than six times as great as in 1841. The knitting of stockings by hand was a domestic industry throughout Derbyshire before the invention of the stocking frames. In 1758 Jedediah Strutt, a native of Derbyshire, invented a rib machine for attaching to Lee's Stocking Frame. The stockings knitted on this machine were known as "Derby ribs," and the industry spread rapidly over the midland counties. These frames were often in the homes of the worker, but the factory system has caused the aggregation of the trade in the centres of Derby, Ilkeston, Heanor, and Long Eaton.

Silk was manufactured in Derby in the eighteenth century. The first textile silk-mill in England was erected

Old Silk Mill, Derby

by John Lombe on an island in the river Derwent which was rented from the Corporation. This mill was at work until a few years ago but has since been removed by the Corporation. During the nineteenth century the industry rapidly increased, but after 1861 declined, owing to the competition with France, and in 1901 there were only about 600 persons employed in the county in the manufacture of silk.

The cotton trade, which is now an important industry in Derbyshire, began in 1771, when Richard Arkwright patented the "water-frame" and subsequently made other improvements in the manufacture of cotton. He left Nottingham, where horse-power was used, for Cromford, at which place he utilised the water-power of the river Derwent. The removal of the excise duty on calico in 1774 revolutionised the cotton industry, which spread rapidly in Derbyshire, and many new mills were built at Belper and Milford by the Strutts, who were partners of Arkwright, and at Glossop. The improvement of bleaching and dyeing and the introduction of cotton printing was a natural result of the growth of the cotton trade. The former trades are now carried on near Glossop and Matlock.

Paper is manufactured at Glossop, New Mills, and Hayfield in the Mersey watershed, and at Bonsall and Little Eaton in the Derwent watershed.

Pottery and china manufacture are also important industries. Pottery was made at Duffield during Norman times, and at Dale Abbey and Repton paving tiles were manufactured in the fourteenth century. In the eighteenth century brown pottery was made in many places in Derbyshire, and at the present day pottery works at Codnor Park, Derby, Ripley, Langley Mills, Ilkeston, Swadlincote, and Hartshorne employ a large number of hands.

Derby is specially noted for its china industry. The first china works in Derby were established by Duesbury on the Nottingham Road. At a later date he bought the

Chelsea china factory, and plant at Bow, and afterwards closed these two works and manufactured only in Derby. The china works remained for three generations in the Duesbury family, and after passing through several employers were discontinued in 1845. In 1877 Mr Phillips, who came from the Worcester Royal Porcelain Works, revived the manufacture of china in Derby on a new site, and in 1890 the Company received the honorific title of Royal and became the Derby Royal Crown China Company.

The iron trade is now the most important industry of Derbyshire, with the exception of coal mining (dealt with in the next chapter). After the middle of the seventeenth century the trade in iron increased, but at the close of that century it declined rapidly because of the importation of cheap Swedish iron and the expense of procuring charcoal for the furnaces and forges.

About the middle of the eighteenth century the substitution of coke for charcoal created a great revival of the iron industry, which has been maintained to the present day. From the census returns of 1901 it appears that engineers and machine-makers formed by far the greater proportion of the iron-workers in Derbyshire. The railway men at Derby, however, form an important branch of this trade. The locomotive and carriage works there employ a great number of hands, and according to the census of 1901, over twelve thousand men were employed for the conveyance of persons and goods on railways in the county.

16. Mines and Minerals. (*i*) General— Coal Mines.

The mines and minerals of Derbyshire are so numerous and interesting that several chapters will be required to describe them. The language of commerce uses the terms minerals and metals in a more or less general sense for the materials forming the crust of the earth which have been obtained either from the surface of the ground, from pits or quarries, or from greater depths in mines.

Each year the Home Office publishes reports of its inspectors of mines. These reports contain information about inspections under three sets of Acts of Parliament —the Coal Mines Regulation Acts, the Metalliferous Mines Regulation Acts, and the Quarries Act. Interesting information is given about the minerals worked, the number of people employed, and the quantities of minerals obtained, and it is convenient therefore to divide our subject into three portions—the coal mines and minerals connected with them, the metalliferous mines, and the quarries ; though lead-mining, being such an old and important industry, will have a separate chapter.

In the year 1906, according to Mr Stokes' report, the number of people employed in Derbyshire in coal mines was 51,904, in metalliferous mines 482, and in quarries 3927. From the coal mines 16,647,224 tons were raised, from the metalliferous mines 46,194, and from quarries 2,579,840 tons.

Coal is the most important and most largely worked

mineral in Derbyshire. It is obtained from three separate coalfields—the North Derbyshire, which is part of the Lancashire coalfield; the Leicestershire and South Derbyshire; and the Yorkshire, Nottinghamshire, and Derbyshire coalfield, on the east of the latter county. Although the presence of coal near to the ironstone must have been known at an early date, we have no evidence to show how early coal was worked in our county. From a charter of Edward II we learn that in 1315 the coals from Derbyshire were used in the monasteries, and the monks of Beauchief Abbey were supplied with coals from mines near Alfreton and Norton.

The amount of coal raised in Derbyshire has increased rapidly each year. Since the year 1808 when a little over a quarter of a million tons were carried by the Cromford, Erewash, and Nottingham Canal, the output has increased, as we have seen, to more than sixteen and a half million tons in the year 1906. There are about 176 coal mines in Derbyshire, and of the 52,000 people employed in them slightly over 41,000 worked underground, and only between ten and eleven thousand above ground. Females have not been employed in Derbyshire coalpits, but before the middle of the nineteenth century boys as young as five years of age were employed for driving the donkeys and hauling the baskets of coal. The Mines Act of 1872 prohibited the employment of very young children in mines, and led to improvements in the methods of mining.

The coal is worked at varying depths in Derbyshire, from 100 to as much as 1700 feet below the surface of the ground. In 1899 about as much coal was raised from the last 600 feet as from the first 400 feet, and by far the greatest quantity was obtained at a depth of between 500 and 1000 feet. In some places the coal is worked at the surface, but in Derbyshire only 4759 tons were obtained from quarries in 1906.

There are several important minerals which occur in the coal-measures in Derbyshire, and are worked in the collieries. These are ironstone, fireclay, gannister, and pyrites.

The Derbyshire ironstone is what is commonly called clay ironstone, and is found nearly throughout the whole depth of the coal-measures in the eastern coalfield of the county. Formerly a large quantity was raised in Derbyshire, but Northamptonshire ore, which is much easier to get, is now largely imported into Derbyshire. In 1906, only 5485 tons of ironstone were raised in Derbyshire, against 56,874 raised in Northamptonshire.

A large quantity of fireclay is obtained from the Derbyshire coalfield. This clay contains little or no iron, lime, or alkali, and will stand intense heat without melting. In 1906, 72,389 tons were raised in Derbyshire.

Gannister, a hard and fine grained sandstone consisting of silica, sometimes forms the floors of the seams of coal in the lower measures. It is used in some places for road-metal and when ground down and mixed with fireclay makes excellent fire-bricks, or forms fire-resisting lining for the inside of furnaces. In 1906, 367 tons were obtained from mines and 1100 tons from quarries.

Pyrites, a mineral composed of sulphur and iron, is frequently found in nodules and in thin layers in coal seams. It deteriorates the value of the coal in which it occurs, and gives much trouble in sorting the coal for iron smelting, but it is used largely for the production of sulphuric acid. Some 1774 tons of it were raised in Derbyshire in 1906.

17. Mines and Minerals. (*ii*) Metalliferous Mines.

Of the 482 persons employed in metalliferous mines in Derbyshire in 1906, the report states that 297 worked below ground and 185 above ground. The quantity of material raised was 46,194 tons, and we may divide the material into two classes, firstly that consisting of gypsum, chert, ochre, and umber, which as commercial products are never found in association with lead ore; and secondly the minerals such as barytes, calc spar, and fluor spar, which often form the matrix enclosing a lead vein ; and zinc ore, which is often found in lead-mines.

Alabaster or gypsum (sulphate of lime) is said to have been raised from Chellaston as early as the fourteenth century. It is obtained by sinking shafts or driving levels into the ground (i.e. horizontal tunnels) and then cutting out in headings, from which the blocks of gypsum are taken out. Though it is now mainly used for making plaster of Paris, the whiter variety called alabaster was used for ornamental purposes. The pulpit and entrance to the choir in St Luke's Church, Derby, and tombs in the churches at Bakewell and Ashbourne are wrought in alabaster. Plaster of Paris is made by baking the gypsum in ovens, and thus evaporating what is called its water of crystallisation, the mineral falling into a white powder. The powder is used for statues, medallions, casts of all sorts, paper-glazing, and many other purposes. The output from Derbyshire in 1906 was 7381 tons.

The mining of chert is a characteristic Derbyshire industry. The chert of commerce is a silicious limestone used in making china and porcelain which is found in thick beds in the neighbourhood of Bakewell. Its crystalline structure differentiates it from the nodules and layers of chert in the upper beds of Mountain Limestone. The chert is obtained in large blocks by taking away the limestone underneath the bed of chert and letting the latter fall by its own weight. In 1906, 3612 tons of chert were obtained from mines and 150 tons from quarries in Derbyshire. A large quantity is sent to the potteries in Staffordshire.

Umber, ochre, and manganese "black wad" are used in the manufacture of paints. In 1906, 74 tons of ochre from mines, in addition to 110 tons from quarries; and 63 tons of umber from mines were obtained in Derbyshire. These substances are found in cavities in the limestone into which they have at some time been introduced by water.

Barytes, locally termed "cauk," is a heavy mineral often found in lead-mines. Formerly this mineral was thrown away on to the old hillocks surrounding the mines. Of late years, since its commercial value has been recognised, it has been obtained from the old tip heaps. It is largely used for adulterating white lead, and probably enters into the composition of paint. It is also used for coating papers upon which impressions from "process" blocks are made. In 1906, 326 tons of barytes were raised in Derbyshire.

Calc spar, or crystallised carbonate of lime, is chiefly found associated with lead ore. Some large specimens have been used for ornamental purposes, but it is chiefly used for garden walks. For this purpose it is broken up and sifted. In 1906, 1298 tons were obtained from mines and 700 tons from quarries.

Fluor spar, or fluoride of lime, is sometimes called fluxing spar, because the commoner varieties are used in smelting copper ores and in plate-glass making. "Blue John" is the name given to the dark purple concretionary spar found in the Blue John mine, and it is much used for making ornaments, which fetch a good price. Fluor spar is used for making one of the most penetrating and corroding acids known, called fluoric acid. This has the property of corroding glass, and is kept in leaden vessels, on which it has no action.

In working the old lead mines, fluor spar and other minerals were thrown aside as useless, so that in the old hillocks or refuse heaps there are large quantities of fluor spar. During the last few years the value of the fluor spar refuse has been discovered, and large quantities have been sent away for fluxing purposes to various parts of England and America. In 1906, 26,984 tons were obtained from mines or mine hillocks, and 700 tons from quarries.

The amount of zinc ore raised in Derbyshire was only 766 tons in 1906. Little attention to the ores of zinc was paid until the eighteenth century. They are often associated with lead ore.

18. Mines and Minerals. (*iii*) Quarries.

Limestone or carbonate of lime is the most widely distributed mineral in Derbyshire. The quantity quarried in 1906 was more than twice as great as that of all the other minerals taken together, and amounted to 1,841,875 tons. It covers a large area of ground in the northern part of the county from Castleton in the north to near Ashbourne and Wirksworth in the south, and is associated with some of the finest scenery. It is used in the manufacture of iron, chemicals, and lime, as a metal for roads, and for building purposes. A large quantity of limestone is burnt at Buxton, Ambergate, and Small Dale near Peak Forest Station. The stone is broken up and burnt in special kilns. The carbonic acid which it contains is driven off and the resulting product is quicklime.

The limestone in the eastern part of Derbyshire is of a different kind and contains magnesia as well as lime. The Houses of Parliament are partly built of magnesian limestone from near Bolsover.

The sandstones of Derbyshire vary very considerably in colour and texture. They are used for paving, building, and roofing, and for making millstones. Some of the

millstone grit is used for engine beds, and other foundations where strength and weight are required. This
rock has been used largely for buildings. St George's
Hall Liverpool, Chatsworth House, and Buxton Crescent,
are built of Derbyshire gritstone.

In addition to the above-mentioned minerals, large
quantities of clay, brick-earth, marl, and shale are
quarried, as well as gravel and sand.

19. Mines and Minerals. (*iv*) Lead and Lead Mining.

The oldest industry in Derbyshire is that of lead-
mining and smelting.

The majority of mines have been worked out, or
abandoned owing to the difficulty of getting rid of water,
the expense of obtaining the ore, and the great fall in the
price of lead. The only mine at which any quantity of
lead is being raised at the present day is the Mill Close,
near Darley Dale, in the upper beds of the Mountain
Limestone. The large number of old mine-heaps or
hillocks and of old shafts bear witness to the vast amount
of lead-mining which has been done in Derbyshire.

The discovery of pigs or blocks of smelted lead with
Latin inscriptions proves that lead ore was raised and
smelted in Derbyshire during the Roman occupation of
Britain.

Since the Roman invasion some of the mines appear
to have belonged to various religious houses, and became

the property of the Crown at a very early period. The lead-mining industry in Derbyshire is governed by curious customs and rights which have existed for many centuries. A poem composed by Edward Manlove, printed in 1653, and a book called *The Articles and Customs of the King's Field in the High Peak of Derbyshire*, published in 1601, give in detail the old customs and rights of lead-mining. These mining rights were confirmed by Acts of Parliament passed in the years 1851 and 1852.

In certain parts of the county anyone may search or dig for lead without asking for the permission of the owner or occupier of the land, and the latter cannot claim compensation. This is subject to the condition that the miner finds lead ore and pays a dish of lead to the barmaster. The miner is entitled to sufficient surface on which to deposit his hillock of waste material, a way to the highway or road most convenient from the mine, and a right of waterway to the nearest stream of running water. The only satisfaction the owner of the land gets for annoyance and loss is the right to sell any other mineral except lead which the miner may bring to the surface.

The miner had to pay dues to the Crown, the Duchy of Lancaster, the barmaster, and in some places to the church of the district. The royalty to the Crown was a certain rate per dish, a dish containing about 472 cubic inches. The barmaster even at the present day carries his dish with him to measure the ore. The standard dish for the wapentake of Wirksworth is of brass, and is kept at the Moot Hall, Wirksworth.

The lead ore chiefly worked is galena or sulphide of lead, which contains a small quantity of silver (two to four ounces per ton). The ore occurs in veins known to the miners as "rakes," "pipes," and "flats." A rake vein is generally an almost vertical fissure or crack in the limestone, and "scrins" are strings of ore which branch off from the rake and form smaller veins. The ore occurs in ribs with layers of calcite or fluor arranged more or less parallel to the walls of the rake or vein. Pipe veins are irregularly-shaped hollows or pockets in the limestone, generally parallel to the bedding-planes and often connected with one another by a crack filled with clay or spar called a leader. A flat is not so common as the rake or pipe veins. It is generally found along the junction of two beds of limestone, and consists of a low flat chamber with the roof and floor only a few feet apart : it seldom has any leaders connected with it.

We can readily understand from the chapters on caverns and underground streams that the miner must often have met with volumes of water greater than pumping engines could cope with. Before steam-engines were introduced, the miners drove adits through the limestone with their mouth on a river or brook-side, so that water running out found its way into the natural drainage of the district. These artificial underground channels for carrying away the water from the mines are called "soughs." The Hill Car and Meerbrook soughs are the longest in the county. The former, near Youl-greave, is about four miles in length, took 21 years to drive, and cost upwards of £50,000. The Meerbrook sough, which drains the Wirksworth lead-mines and empties the water into the Derwent near Whatstandwell, was commenced in 1773, is three miles in length, and cost £45,000.

With few exceptions the descent into the mines is by ladders, "stempels," and footholes. Stempels are pieces of wood, about four to six inches in diameter, fixed from one side of the shaft to the other and fastened at each end in the rock. They are placed a few inches from the side and about two feet apart on opposite sides of the shaft. The miner descends by planting one foot on a stempel on each side of the shaft, and moving the feet alternately on to lower stempels. The most dangerous way of climbing a shaft is by footholes, which are small holes about eighteen inches apart made on opposite sides of a narrow shaft. The miner places his feet in them with legs astride in the shaft, his back pressing against the side of the shaft, and in descending places the palms of his hands where his feet have been, taking alternate steps until he reaches the bottom of the shaft.

20. History of Derbyshire.

In considering the present-day condition of a country we must by no means omit a review of its past, for the one is but the outcome of the other and is indissolubly connected with it. The complex social and political life of the present day is a growth from small and seemingly unimportant happenings perhaps hundreds of years ago, and it is the tracing of these which makes history so interesting. Let us now see what these early influences in Derbyshire have been.

In Chapter 13 we have already sketched the history of the county up to the time of the Norman Conquest. Soon after the Normans settled in England a most interesting and valuable survey and series of statistics were made and recorded in the *Domesday Book*. The portion dealing with Derbyshire is short, but shows how much the county has been influenced by Scandinavian rule. The population of Derbyshire was 2868, including the exceptionally large number of 42 censarii, i.e. men who paid money rent instead of service.

Derbyshire was one of the very few English counties which had a distinctive industry at the date of the survey. Its lead-mining was evidently important, though the greater part of the county was agricultural. The King was the chief holder of lands here as elsewhere, and these were grouped into large manorial blocks for the sake of agricultural organisation. He owned a series of manors stretching from Ashbourne to the York border, and the payment from these before the Conquest had been in " honey and lead." Ashbourne is the only recorded town paying in "pure silver." William de Ferrers was custodian of these manors for a time, but soon acquired possession of them.

William the Conqueror also owned the forfeited estates of Edwin, late Earl of the Shire, and his eight messuages in Derby, the valuable manor of Melbourne, and a group of manors with their satellite " berewicks."

Berewicks were settlements connected with barns for the collection of corn, a "wick" being a village in which barley was grown.

The growth of towns was accelerated under the royal ownership of land, as it was easier to obtain privileges from the distant King, who was always wanting money for his wars and national improvements, than from the wealthy neighbouring barons.

The King granted large portions of his land to the different earls in fee-farms for life. Henry de Ferrers at one time owned one hundred and fourteen manors in the county, and a large part of the political history of Derbyshire is concerned with his house. His chief castle was at Tutbury just outside the county boundary, and he founded a Priory there before 1086. The chief credit of the victory of the Battle of the Standard at Northallerton against the Scots is due to his son, Robert de Ferrers, who obtained the title of Earl of Derby as a reward.

The Peverils of North Derbyshire also played an important part in history, and William Peveril's castle of Peak is mentioned in the Domesday record, but it passed to the crown in the reign of Henry II.

Derbyshire took little part in the wars of the Barons, and the three fortresses of Castleton, Bolsover, and Horsley were generally held for the King, who appointed William de Ferrers, son and heir of the insurgent baron who lost his castles in 1174, as custodian. King John visited his Derbyshire castles several times, and rebuilt Horsley. Earl Ferrers took arms against King Henry III, and all his lands were confiscated, and his local possessions conferred on the King's son Edmund, Duke of Lancaster. Some of this property still remains with the royal "Duchy of Lancaster." It was then, probably, that Duffield Castle was demolished.

Derbyshire was linked with Nottinghamshire for civil administration up to Henry III's reign; the assizes were held only at Nottingham, and from the time of Henry III. to that of Elizabeth they were held alternately at Derby and Nottingham. There was only one county gaol at Nottingham for the two counties.

Edward I's reign marks the beginning of parliamentary rule, but no Derbyshire names are mentioned in parliamentary lists until 1295. Grants of markets were made to a number of Derbyshire towns during the thirteenth and fourteenth centuries, among them Ashbourne, Wirksworth, Melbourne, Sandiacre, Bakewell, Monyash, and Ilkeston. The chief industries were wool, wine, and lead, and a good deal of trade was done in these with the rest of England and with the Continent. Fulling and dyeing were carried on in Derby, and the name Full Street still survives.

The Third Derby Charter was obtained in 1204; it enacted that no cloth should be dyed within a radius of ten leagues from Derby except at Nottingham, it also allowed the burgesses to form guilds. The guilds were very important during the middle ages and exercised a great monopoly in trading, though they were themselves sometimes fined for taking excessive tolls. Although they might oppress individuals yet they stood up for the liberties of the town against any infringement of them by the King, and against foreigners. The history of the wool trade in Derby is a long series of disputes between the King, the Abbots of Darley, and the burgesses.

The earliest trading was private and retail, but it developed into municipal trading when civil officials were allowed to buy and sell in the various markets.

There was little besides architecture and dress in which money might be invested at this time, so money accumulated, and the guilds often became very wealthy bodies. Besides carrying on trade, they built churches, repaired and built bridges and town walls, and assisted the poor. Chesterfield had several guilds in the thirteenth century, and there also was one at Dronfield and another at Tideswell.

Derbyshire was closely connected with John Baliol, the claimant for the Scotch throne in 1291. He held the custody of the Peak and the honour of Peveril, and served as Sheriff for Derbyshire and Nottinghamshire from 1261–4. All the leading men were engaged in his wars until his deposition in 1296.

In 1327 Queen Isabella held the High Peak Castle, town, honour, and forests of the then yearly value of £291. 13s. 4d., which sum must be multiplied by fifteen to bring it up to the present value of money.

During the fourteenth century the archers of Derbyshire were of great repute, and frequent levies were made for the wars with France, Scotland, and Spain. One hundred and fifty bowmen were provided by this county for Agincourt in 1415. Edward III commissioned 500 archers and 200 hobelers (light horsemen) to fight against the

Scots, but there were frequent desertions and punishment. The archers continued to be famous, and 4510 of them are included in the muster of April, 1539. These county musters played an important part in Elizabeth's reign and onward until James I abolished the old form of military service.

There was great dissatisfaction felt in Elizabeth's reign because some of the miners were compelled to serve in quelling the rebellion in Ireland, while the mines were standing idle and the men themselves were quite untrained for service.

The Black Death was very severe in Derbyshire in 1349 and the plague again visited the county in the sixteenth and seventeenth centuries. Grass grew in the streets of Derby, and the Hedles Cross in the Arboretum once stood on the western outskirts of the town to mark the spot where temporary fairs or markets might be held with least danger of infection. During these terrible visitations of plague harvests were ungathered, and many deaths occurred amongst all classes of the people. The story of the heroic action of Mompesson, the vicar of Eyam, in his efforts to prevent the spread of the disease and to help his stricken people is well known.

There was much religious persecution during the sixteenth and seventeenth centuries and the wild places in the Peak formed convenient hiding places for fugitives. Several important Derbyshire families were Catholic, as owing to the comparative inaccessibility of this county and the hill country generally, the old faith lingered on more effectively than in the more civilised south. The last Protestant martyr in Derby was Joan Waste, a girl only 22 years of age, who was burnt for her faith at Windmill Pit in 1556.

Derbyshire, probably owing to its inaccessibility and its distance from the coast, is celebrated for its illustrious prisoners. John of Bourbon was taken prisoner at Agincourt, and confined at Melbourne Castle for nineteen years.

Mary Queen of Scots spent much of her unhappy captivity in Derbyshire at Chatsworth, Derby, Buxton, and Wingfield, until her final custody at Tutbury Castle, preceding her execution at Fotheringay in 1585.

Anthony Babington, the leader of the conspiracy to murder Elizabeth and to set Mary Queen of Scots on the English throne, owned a house in Derby in Babington Lane, where Queen Mary stayed in January, 1585. The plot was discovered by Walsingham, and Babington was executed in 1587.

In the seventeenth century a large number of Scotch prisoners were confined in the church at Chapel-en-le-Frith, and in sixteen days 44 had perished from cold and starvation.

A number of French prisoners were sent to Derbyshire during the wars of the eighteenth and nineteenth centuries and the art of netted glove making was introduced by them into Chesterfield.

George Fox, the celebrated founder of " The Society of Friends," was also a prisoner in Derby for a year, " amongst thirty felons in a close stinking place." At his trial the name " Quaker " was first given to him by

Justice Bennett, in allusion to the tremblings that formed part of his ritual in preaching. Derby was also the first place where a female quaker preached.

James I visited Derby, and King Charles also visited the town, staying three days at the "Great House" in the market place, and obtaining £300 from the Corporation.

During the civil wars many Derbyshire men fought against the King under Sir John Gell of Hopton, who had previously made himself notorious over the collection of ship-money.

The Royalists were defeated at Swarkestone Bridge and at Ashby-de-la-Zouch. Gell was made Governor of Derby and kept his main guard in the Town Hall for four years. His soldiers were similar to the levies raised at the beginning of the war, "a set of poor tapsters and town apprentices" as Cromwell calls them. The town and the neighbouring gentry grumbled at the expense of keeping them so long, as it exceeded that of other towns by three thousand pounds. Farmers, moreover, were afraid of bringing their produce to Derby market lest they should be robbed by the soldiers.

Gell was successful in the siege of Lichfield Cathedral, and again at the battle of Hopton Heath near Stafford, where the Earl of Northampton was killed. The body was brought to Derby for burial in the Cavendish vault in All Saints' Church. The Royalists pillaged up to the gates of Derby and stormed Wingfield Castle, but Gell regained it after one month's siege. The county was in a miserable state during its varying fortunes, and in 1646 there was a serious mutiny in the Derbyshire regiments. The unpopular and tactless Secretary of State to Charles I, Sir John Coke of Melbourne, was a Derbyshire man.

William Cavendish, fourth Earl of Devonshire, played a great part in driving James II from the throne, and bringing William of Orange over to England. The final conspiracy was hatched in a small house on Whittington Moor, near Chesterfield, and its restored remains are still called "Revolution House." William Cavendish was made a Duke by William III as a reward for his services.

In August, 1709, the whole county was thrown into a ferment by an assize sermon preached by Dr Sacheverell at All Saints' Church, Derby. It was a covert attack on the revolution of 1688, and advocated the principle of non-resistance to supreme power. After a nine days' trial, he received a mild sentence which the Tories regarded as an acquittal, whereat bells were rung and bonfires lighted in Derby as tokens of joy.

In 1745 Swarkestone Bridge became famous as the most southerly point reached by the pretender Charles Edward and his force of seven thousand men. The more important men of the county joined the tradesmen and yeomen to resist the "Popish Pretender," but no battle was fought, as the officers turned faint-hearted, and Charles Edward, much to his regret, was obliged to retreat with his force.

Soon after the Restoration a post was established twice a week between Derby and London, and in 1719 the first Derby newspaper made its appearance.

The county was in a very unsettled state from the time of the French Revolution to the passing of the Reform Bill. There were riots at Bakewell, Ashbourne, and Wirksworth, due to the rebellion of the miners against forced service in the militia. In 1811 the Luddite riots commenced and the stocking-frame breakers did much damage. The distress amongst the working classes reached a crisis in 1817. A murderous scheme was hatched, incited chiefly by those in authority. It was soon quelled but, for the sake of example, two stone-masons and one frame-knitter from Pentrich suffered the outrageous death of being hanged, drawn, and beheaded (instead of being quartered) for " High Treason." Shelley the poet was in the crowd and commented unfavourably upon the execution.

21. Antiquities.

The earliest history of the people who inhabited Derbyshire is not derived from written records but from the relics or antiquities which have been found in the county.

Archaeologists divide the prehistoric period during which Great Britain was inhabited into four more or less distinct periods. These are the Palaeolithic (old stone) period during which man used rude implements and weapons of stone ; the Neolithic (new stone) period when he used more highly finished weapons of stone but was still ignorant of the use of metals ; the Bronze age when his implements were made of copper and tin ; and the Iron age when the use of that metal became established.

Relics of man of the Old Stone period are found in caverns and old river gravels associated with bones of animals which are now extinct in Britain, whilst those of the New Stone age are found associated with wild and domesticated animals, many of which are similar to those now living in Great Britain. The break between these two stone ages is of a very marked character and there is little doubt that a very lengthy period of time separated them. During the Neolithic or New Stone age, the climate and general surface of the country were very little different from what they are to-day, whilst the climate of the Palaeolithic or Old Stone age was so much colder that a large portion of Great Britain was coated with ice. During this age, Great Britain was united to the Continent of Europe.

The Recent age of the Geologist includes all prehistoric time from the Neolithic, through the Bronze and Iron ages, to the present time. The changes from the Neolithic age to the Bronze and Iron ages show no gap like that between the two Stone ages, but were gradual. The different materials doubtless remained in use side by side for some time, and the change from one period to another would be made at different times in different districts.

The antiquities subsequent to the Iron age we generally classify as Roman and Saxon, corresponding with the history of our land from 55 B.C. to 1066 A.D.

The only traces of Palaeolithic man found in Derbyshire occurred in the Creswell Caves on the N.E. border of the county. These caves have yielded results only surpassed in England by those of Kent's Cavern at Torquay. In 1876, no less than 2726 bones and 1040

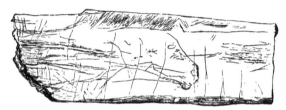

Fragment of rib with engraving of horse (*full size*), Robin Hood Cave

Bone needle, Church Hole Cave (*full size*)

Bone awl, Church Hole Cave (*full size*)

implements were obtained from Robin Hood Cave, and 1604 bones and 234 implements from Church Hole Cave. The implements were of quartzite, flint, and bone. Amongst other interesting implements found in Church Hole Cave were "a well-shaped needle, absolutely perfect, made out of a metacarpal or tarsal bone of a ruminant and much larger than any of those figured from the Palaeolithic caves of France, Belgium, or Switzerland" and "two bone awls fashioned out of the tibia of a hare and polished by long-continued use." On one flat piece of bone was scratched a sketch of the head and fore-quarters of a horse. This was the first trace of pictorial art discovered in Great Britain. The majority of the bones were those of animals now extinct but alive in this country when it was part of the Continent of Europe during Palaeolithic times. The similarity of these de-posits to those in France and Switzerland proves that the hunters of those countries found their way along the Eastern Valley, now covered by the German Ocean, and wandered as far north as southern Yorkshire.

The remains of Neolithic, Bronze, and early Iron man in Derbyshire consist mainly of burial mounds and other remains of tombs. Nearly 300 of these have been opened, and their contents examined scientifically, but many others must have been destroyed and used for building and other purposes. The common word in Derbyshire for a burial ground is "low," from the Saxon word meaning a small hill or heap, and it often occurs in place names. These burial mounds have been divided into three kinds, those which belong to the Neolithic stage, those of the Bronze age, and those of later type which may belong to Roman times.

There are about a dozen structures in Derbyshire popularly known as druidical circles. The two finest are at Arborlow, near Parsley Hay station, on the

L. & N. W. Railway between Buxton and Ashbourne ; and the "Bull Ring" at Dove Holes, on the L. & N. W. Railway, between Buxton and Manchester. It is considered that they were connected with religious rites, and, as Sir Norman Lockyer has shown, they were probably erected as rough astronomical instruments.

There are in Derbyshire ten or twelve defensive works or fortifications which are not of Roman origin.

Stone Circle, Arborlow

They are probably pre-Roman and were used as refuges in times of tribal insecurity. The largest of these are on Mam Tor near Castleton, Carl's Wark near Hathersage, and the fort at the western end of Combs' Moss, near Dove Holes.

Derbyshire contains interesting traces of the Roman occupation. The three Roman forts in Derbyshire were Melandra, Anavio or Brough, and Little Chester. Melandra Fort is near the borders of Cheshire, and not far from Dinting Vale station on a branch of the Great Central Railway. It lies near the confluence of the Glossop brook and the Etheroe. It commanded the easy access to the hills of North Derbyshire. Brough Fort is near Hope station on the Dore and Chinley branch of the Midland Railway. It is near the confluence of the Bradwell Brook with the river Noe, and commanded the

Mam Tor, Castleton

approach to the hills in the neighbourhood of Castleton and Bradwell. Little Chester is in the N.E. part of the present borough of Derby. It is situated as other Roman forts often are, in an open valley, close to water, and commands the approaches to the hilly country to the north.

The Roman Road in Derbyshire known as Ryknield

Street passed through Lichfield, Little Chester, and Chesterfield to Templeborough near Rotherham. Four other Roman roads branched out from Little Chester. One led to Buxton (called by the Romans *Aquae*), probably a Roman village, a shorter road ran to Sawley-on-Trent, and another may have led westwards to Rocester on the Staffordshire border. From Melandra, the road known as the Doctor Gate led across the moors eastward to Brough and ultimately to Templeborough, another led to Stockport and Manchester, and another possibly to Buxton.

From Brough the Batham Gate ran up the Bradwell valley and across the moors to Buxton ; and Doctor Gate and Long Causeway connected Melandra, Brough, and Templeborough.

In the neighbourhood of Buxton and Wirksworth Roman remains have been found in caverns showing that they were inhabited during the Roman occupation. Numerous other small finds are scattered over the hilly country, which point to temporary occupation only. Roman "pigs" of lead have been found near Matlock and Brough, which show that the lead-mining industry flourished in Roman times.

The Saxon remains in Derbyshire, which consist of grave-mounds, earthworks and other visible monuments, are few compared with those of prehistoric age. A number of barrows or mounds, however, have yielded ornaments and utensils which sometimes are found with the skeletons or bones of uncremated bodies, and in other cases with the calcined bones or ashes.

22. Architecture—(a) Ecclesiastical. Churches and Crosses.

Dr Cox remarks in his exhaustive work on the *Churches of Derbyshire* that "this county cannot vie with Somersetshire in its towers, with Northamptonshire in its spires, with Norfolk or Suffolk in the size or beauty of so many of their churches, or with Kent in the number of its brasses ; but no other county of the same size has anything like the same extensive variety of styles, and excellent specimens of every period, both in the ecclesiastical fabrics themselves and in the monumental remains." So that, although Derbyshire possesses no cathedral like Lichfield or Southwell, much may be learned from a study of its "history in stone." A continuous development may be traced from the Romanesque period on through the various Gothic periods ; each one embodying the ideals and aspirations of the people, and never copying the art of a past age. But there was no abrupt change in any one year, at all places alike, for styles nearly always overlap, and progress was not at all uniform in different parts of the country.

A preliminary word on the various styles of English architecture is necessary before we consider the churches and other important buildings of our county.

Pre-Norman or—as it is usually, though with no great certainty termed—Saxon building in England was the work of early craftsmen with an imperfect knowledge of stone construction, who commonly used rough rubble walls, no buttresses, small semi-circular or triangular arches, and square towers with what is termed "long-and-short work" at the quoins or corners. It survives almost solely in portions of small churches.

The Norman Conquest started a widespread building of massive churches and castles in the continental style called Romanesque, which in England has got the name of "Norman." They had walls of great thickness, semi-circular vaults, round-headed doors and windows, and massive square towers.

From 1150 to 1200 the building became lighter, the arches pointed, and there was perfected the science of vaulting, by which the weight is brought upon piers and buttresses. This method of building, the "Gothic," originated from the endeavour to cover the widest and loftiest areas with the greatest economy of stone. The first English Gothic, called "Early English," from about 1180 to 1250, is characterised by slender piers (commonly of marble), lofty pointed vaults, and long, narrow, lancet-headed windows. After 1250 the windows became broader, divided up, and ornamented by patterns of tracery, while in the vault the ribs were multiplied. The greatest elegance of English Gothic was reached from 1260 to 1290, at which date English sculpture was at its highest, and art in painting, coloured glass making, and general craftsmanship at its zenith.

After 1300 the structure of stone buildings began to be overlaid with ornament, the window tracery and vault ribs were of intricate patterns, the pinnacles and spires loaded with crocket and ornament. This later style is

known as "Decorated," and came to an end with the
Black Death, which stopped all building for a time.

With the changed conditions of life the type of
building changed. With curious uniformity and quick-
ness the style called "Perpendicular"—which is unknown
abroad—developed after 1360 in all parts of England and
lasted with scarcely any change up to 1520. As its name
implies, it is characterised by the perpendicular arrangement of the tracery and panels on walls and in windows,
and it is also distinguished by the flattened arches and the
square arrangement of the mouldings over them, by the
elaborate vault-traceries (especially fan-vaulting), and by
the use of flat roofs and towers without spires.

The mediaeval styles in England ended with the
dissolution of the monasteries (1530–1540), for the
Reformation checked the building of churches. There
succeeded the building of manor-houses, in which the
style called "Tudor" arose—distinguished by flat-headed
windows, level ceilings, and panelled rooms. The ornaments of classic style were introduced under the influences
of Renaissance sculpture and distinguish the "Jacobean"
style, so called after James I. About this time the pro-
fessional architect arose. Hitherto, building had been
entirely in the hands of the builder and the craftsman.

Most of the Saxon churches have long since perished,
but round chancel-arches of probable Saxon work may be
seen at Marston Montgomery, Sawley, Stanton-by-Dale,
and Ault Hucknall, this last-mentioned church containing
much work in Saxon style. There is a very remarkable
Saxon crypt of late eleventh century date under the altar

at Repton, with spirally twisted pillars and a vaulted roof, which is unique in Britain.   A characteristic tri-angular-headed window, i.e. two stones inclined at an angle, may be seen at Sandiacre, and the sides built in

Saxon Crypt, Repton

"long and short" work, i.e. alternate bands of stone of different widths.   At St Chad's Church, Wilne, there is a Saxon font of eighth or ninth century date, possibly the oldest in England.

Derbyshire is particularly fortunate in possessing two such perfect examples of Norman or Romanesque work as the large parish church of Melbourne, near Derby, and the little gem of late Norman work at Steetley Chapel in the extreme north-east of the county.   St Michael's Church, Melbourne, is built on the cruciform plan, and has a central tower and a fine west porch flanked by two small towers.

Saxon Font, St Chad's, Wilne

At the east end, the spring of the original three semi-circular apses of the chancel and transepts may be plainly seen.   The thick walls and huge circular pillars on their square bases give the idea of strength and permanence so characteristic of our Norman conquerors.   These early builders knew nothing of the laws of weight and thrust, and so constructed every part strong enough to carry the upper walls and roof. The round arches of the arcades and chancel and the small round-headed clerestory windows on the north side are of typical Norman design.

Steetley Chapel had been used as a barn for 350 years, but divine service was once more held there in 1875. It has an unique richly-ornamented apse at the east end,

St Michael's, Melbourne

with a decorated stone vault, and is ornamented on the outside with shallow buttresses. Many Norman ornaments may be studied here, including beak-heads, cones, billets, zig-zags, grotesques, and foliage on the capitals and arches.

At Whitwell there is a fine cruciform Norman church

considerably altered in the fourteenth century, and other interesting remains of Norman work may be seen at Aston-on-Trent, Bakewell, Longford, Sandiacre, and Youlgreave. There are fine south doorways at Allestree,

The Nave, St Michael's, Melbourne

Breadsall, and Long Eaton, and particularly at Bradbourne. Norman fonts still exist at Mellor and Tissington, and a lead one at Ashover. There is a strange font at Youlgreave with what appears to be a stoup for holy-water attached.

The reign of Henry III marks the age of national development, and a freer and more artistic spirit is shown in this Early English period of Gothic architecture.

In its arcade, chancel, and the three sedilia Ilkeston Church shows the Transitional period, which employed new methods of construction coupled with old Romanesque ornament. Ashbourne Church was built about the middle of the thirteenth century, and the choir is one of the best examples in the kingdom of the grace and beauty of the style. There are two triple lancet windows in the north transept. There is also a fine doorway with characteristic "tooth" moulding. Other details such as clustered pillars and pointed arches may be seen at Wirksworth, and typical lancet windows at Stanton-by-Dale, Doveridge chancel, and Weston.

Early English towers occur at Breadsall and Eckington, but the spire at Breadsall is of later date. "Broached spires" of this or later periods may be seen at Ockbrook, Bolsover, Sandiacre, and Horsley.

A great number of Derbyshire churches seem to have been practically rebuilt in the Decorated Period, viz. 1250–1350 A.D., and we have therefore a number of particularly fine examples.

An important characteristic of this period is the development of tracery in windows. The grouping of lancet windows under an enclosing arch left a space in the head that was first pierced with geometrical forms, such as circles and trefoils, which later developed into beautiful bar tracery, becoming more and more twisted and contorted into meaningless forms, though never descending quite into the excessive ornamentation of the flamboyant tracery of the Continent.

Examples of Decorated windows of different dates and other details can be seen at Sawley, Weston-on-Trent, Norbury, Hathersage, Tideswell, Chesterfield, Ashbourne, Bonsall, Crich, Chaddesden, Ilkeston, Sandi-

St John the Baptist, Tideswell

acre, and Wilne, some of the windows of the last two churches being almost flamboyant in character.

St John the Baptist, Tideswell, is a particularly fine church of grand proportions of the Decorated Period, situated in a little village some $2\frac{1}{2}$ miles from Miller's Dale. It has many important features, including an unusually large chancel, three fine stone sedilia, a stone sepulchre, a stone reredos, an ancient font, a window with "reticulated" tracery, and a tower of late date (probably about 1380) having a wonderful combination of turrets and pinnacles. The crooked spire of All

All Saints, Chesterfield

Saints Church, Chesterfield, provokes much interest. It was built about 1350, and the twisting is possibly due to the use of unseasoned timber.

Fairs, markets, and revels were sometimes held inside

this church, as elsewhere in mediaeval times, and the sacks of wool stored here formed a hiding place for Robert de Ferrers in 1265.

Ashbourne Church has a fine tower and a very graceful octagonal spire of the Decorated Period called "The Pride of the Peak." Its angles are ribbed by strings of ball flowers, and it possesses twenty windows.

The visitor to the fine church of Hathersage is shown the supposed grave of Little John, which is about ten feet long. His bow and cap formerly hung in the church. Mention should also be made of the beautiful chancels of Dronfield, Norbury, and Sandiacre.

The love of ornament and of sculptured forms threatened to run wild in Decorated times, but all building was summarily checked about the middle of the fourteenth century, owing to the long war with France, but above all to the Black Death of 1348 and 1349.

When architecture once more asserted itself, it was sobered and disciplined. The Perpendicular style was that of rigid, straight lines in window tracery, and of superficial ornament and panelling, with which everything was overlaid. The construction, however, was good, and the ornament was never used to disguise construction.

Many of the churches had the steeply-pitched roofs lowered and made nearly flat, probably the timbers had perished near the outer walls and were taken out and cut. The weather-line moulding of the old roof can often be seen on the tower as at Ashbourne, Allestree, and Wirksworth. The aisle walls were often raised and small square-headed windows inserted as at Allestree, Sandiacre, and Wilne, etc. Details of the style may be studied at Elvaston, Longford, Youlgreave, and North Wingfield, and also in the beautiful tower at Youlgreave, the fine late Perpendicular tower of All Saints, Derby, the roofs of Longstone and Repton churches, the screen at Chesterfield, the exceptionally fine one at Fenny Bentley, and the unique fourteenth century oak pulpit at Mellor. Perhaps the churches which show the best series of examples of several periods are Ashbourne, Norbury, Sandiacre, Youlgreave, and Wirksworth. North Wingfield Church still boasts its old wooden roof of 1350, and Kirk Langley and Chesterfield have interesting screens of the Decorated Period. Radbourne Church possesses the fifteenth century woodwork from Dale Abbey.

Derbyshire is very rich in monumental remains in stone obtained in the district, and consequently the number of brasses is not so great, though there are some good ones at Morley, Tideswell, and Sawley.

The Celtic crosses in the churchyards of Eyam, Bakewell, Bradbourne, Hope, Blackwell, Norbury, Spondon, and Taddington are interesting. At Bakewell there are over 100 fragments of sculptured stone, none later than 1260, and many earlier than 1100. At Chelmorton and Darley Dale there are also a large number. Many of them are emblematical and represent shears, keys, swords, axe, bugle, chalice, cross, and stars. The axe is often supposed to mark the grave of the village carpenter, but more probably marks that of a knight, or man-at-arms. The knife is rare, it may show the grave of the official "Kerver" of some great family, which was a post of

Eyam Cross

honour.   Early stone effigies may be seen at Darley
Dale, Melbourne, and Youlgreave, and stone lecterns at
Chaddesden, Crich, Etwall, Mickleover, and Spondon.

The sedilia at Dronfield, Ilkeston, Monyash, Sandiacre,
and Whitwell are remarkably good, and most uncommon
and noteworthy stone chancel screens may be seen at
Ilkeston and Chelmorton, and a stone parclose in Darley
Church.

The earliest glass in the county is probably that of
the Early English lancet window in the west wall of the
north transept at Ashbourne. Egginton has some splendid
glass of about 1300 A.D., but unfortunately very little, and
the finest of all, of about fifty years later in date, is to be
seen at Norbury.   The late fifteenth century glass from
Dale Abbey has found a home in Morley Church, and
some fine early seventeenth century is at St Chad's, Wilne.
The east window of Youlgreave Church was filled in
1876 with beautiful glass designed by Burne Jones, and
the south transept window of Darley Church is a famous
design of "The Song of Solomon" by the same accomplished artist.   There is also some good modern glass in
Ashbourne Church, including a "St Cecilia" window.

23.   Architecture—(b) Religious Houses.

The religious houses of Derbyshire, though not
numerous, have an interesting and unique history.   A
large portion of the church lands and many of the
churches and manors belonged to the monasteries of
the neighbouring counties.   Henry I gave the collegiate

All Saints, Bakewell

church of All Saints, Derby, and the church at Wirksworth to the Dean of Lincoln, as his predecessor, William Rufus, had given Ashbourne and Chesterfield. Melbourne Church belonged to the Bishop of Carlisle, who also owned a palace at Melbourne, in which he took refuge when driven from Carlisle by the Scots. The churches of Bakewell, Hope, and Tideswell, from the thirteenth century for 300 years, were a constant source of dispute between the abbot of Lenton Priory and the Chapter of Lichfield. Even now Derbyshire can claim no cathedral of its own, but belongs to the see of Southwell, and until lately was part of the see of Lichfield.

One of the oldest monasteries in the county was at Repton, the capital of Mercia. It was founded early in the seventh century " for religious men and women under the government of an abbess." Several of the Mercian kings were buried at Repton. The monastery was destroyed by the Danes, but in the twelfth century Repton became famous for a Priory of Austin Canons transferred from Calke. It was destroyed by Henry VIII at the time of the dissolution of the monasteries, many of which had become so corrupt that they well deserved their fate.

Repton Priory illustrates the usual plan for erecting a monastery, viz. a square cloister-garth surrounded by various buildings, but as the river Trent flowed on the north side, the church, contrary to custom, had to be built on the south boundary, and the refectory or dining room on the north. The picturesque gateway to the precincts still remains, and the hall is used as a school-room. The Chapter House and the Calefactory or warmed sitting room, with a dormitory above, were on the east, and the kitchens, buttery, and cellars with a guest hall built above them on the west. The other buildings included a "yelyng-house," or brewing house, a "boultyng-house" where the meal was sifted, and a

Repton School

"kyll house," which might have been the slaughter-house, but was probably the kiln-house. Repton School has been erected on the ruins of the priory, and part of the mediaeval brickwork of Thacker's house, erected after the dissolution, is in a good state of preservation, as is also "the Prior's Lodging," an excellent example of early brickwork (1436–38).

Derby School is also on the site of an old religious establishment of Austin Canons, who removed thence to Darley, near Derby, and founded a very important abbey, of which some small domestic remains survive. They owned three Derby churches, besides several in the county, and whenever the vicars of these churches were ill, the abbot and convent were bound to provide them

Derby School, St Helen's House

with food and clothing as long as they lived. The vicars had all fees from their churches except the mortuary fee, which had to go to the over-ruling convent of Darley. When a householder died, his best beast went to the over-lord, and his second-best beast went as a mortuary fee, provided he had more than three beasts. If he left no beasts, then some of his household furniture or wearing apparel might be confiscated as the convent's fee.

A colony of Premonstratensian or White canons was established at Beauchief, near Sheffield, and another at Dale, near Stanley, the site of which is still marked by a fine east window arch of the same type as, and not much later than that, at Lincoln Cathedral. There is an interesting account extant of the foundation of this monastery. A holy baker, of St Mary's Street, Derby, had a vision of

Dale Abbey Church and Guest-house

the Blessed Virgin Mary, who commanded him to go to Deepdale "and there serve my son and me in solitude." He found the place, "a marsh exceeding dreadful," and then in the rock under the mountain he cut out for himself "a very small dwelling, and an altar turned to the south, which is preserved to this day."

Derby had a house of Benedictine nuns in Kings Mead, who were for a time under the control of the powerful Abbot of Darley. There was a flourishing house of Dominican or preaching friars in a street in Derby, which is still called Friar Gate. The Normans established leper hospitals in Derby, Chesterfield, Locko, Alkmonton, and Spital-in-the-Peak, near Hope. The Locko hospital was preserved down to the fourteenth

Dale Abbey Hermitage

century as a preceptory of the semi-military order of Knights of St Lazarus, and was the only one in England.

Derbyshire suffered severely from the Black Death in 1348–9. Two-thirds of her beneficed clergy died, and must have been much missed, for they not only said mass and administered the sacraments, but assisted the poor, and often acted as schoolmasters.

24. Architecture—(c) Military. Castles.

The foundations of an enormous keep at Duffield, four miles from Derby, were accidentally discovered in 1886. The walls are sixteen feet thick, and form a rectangle 95 feet by 93 feet, so the keep must have been larger than almost any in England except the Tower of

Peveril Castle, Castleton

London. Duffield Castle was a very important stronghold of the Ferrers family, and was probably demolished about 1266 after Robert de Ferrers' defeat at Chesterfield in his fight against the King.

We can obtain a good idea of what a Norman keep was like from the ruins of Peveril Castle at Castleton. It presents the minimum of comfort with its three stories, one partly underground, and no fireplace anywhere. Probably the fire was in the middle of the room with a ventilating shaft in the roof. The keep and courtyard were defended on one side by a strong wall, on the other by a precipice. The courtyard gave a little breathing space, and made a place of refuge for cattle and dependents in time of war. Peak Castle was built by William Peveril

Codnor Castle

in the days of the Conqueror, but it passed to the Crown on the forfeiture of Peveril's son's estates. King John visited his two castles of Peveril and Bolsover in 1200, also his castle of Melbourne, and the castle of Horsley in 1209, but little or no remains are left of these Norman castles.

There was also a castle at Codnor assigned to

William Peveril by the Conqueror and afterwards owned by the Greys, of which some portions survive, and Mackworth, the home of the Touchets, has a pretty remaining fragment.

Tutbury Castle, although just outside the county boundary, undoubtedly played a great part in the history of the county.

25. Architecture—(*d*) Domestic.

We have seen how much geology has influenced the geography of our county. It largely determines the natural history, the animals, birds, and flowers, it also affects the landscape, the rivers, and the watersheds, and by them the industries and population of the county. But geology is also a very important factor in determining the houses of the people. In the southern portions of the county, red brick walls with either thatch or red tiles were largely used, as at Repton and Norbury. Specimens of mediaeval brickwork still remain at Thacker's House, Repton. Sometimes "half-timber" was used, as may be seen at Somersall Herbert, Sudbury, Hartshorn, Mickleover, Derby, and Hilton. In the central portion, a mixture of brick and stone was employed, brick walls with stone dressings, as at Sudbury, Derby (including Babington House, and houses in the Wardwick and Friargate), Longford, Duffield and Stydd; while in the northern portion stone only was used, giving a typical Derbyshire style of which Haddon is the supreme

Dovecote, Codnor Castle

The Mayor's Parlour, Derby

example, though hundreds of others might be mentioned, including those at Alport, Bakewell, and Ashford.

The stone domestic buildings are generally simple and well-proportioned, with gables and mullioned windows, and roofed with "grey" or stone slates, or they are sometimes thatched as at Kilburn and Biggin.

Derbyshire has several fine seats and manor houses,

The Peacock, Rowsley

particularly of the sixteenth century and early seventeenth century. Peveril Castle, already described, represents the earliest and most comfortless method of living. Haddon Hall on the other hand gives us a good idea of what feudal life might be when no longer a question of incessant attack or defence. Haddon has two courts with a fine banqueting hall between. In this hall there are still the old raised table and the minstrel's gallery. The kitchens are at one end of the building, the sitting rooms at the other.   The manor is in a good state of preservation, though it has not been used for a dwelling for nearly 200 years. George IV was the last to occupy the state apartments. Sir George Vernon, in the time of Henry VII and VIII, kept a large retinue of servants here, and it was his daughter Dorothy who was supposed to have eloped with Sir John Manners down the moss-covered steps leading from the beautiful ball-room.   The cottage garden has yew trees clipped to represent a boar's head and a peacock, the arms of the Vernon and Manners families. There are architectural examples of five distinct periods, and this, together with its romantic situation, makes Haddon one of the most interesting old buildings in England.

Wingfield Manor is the stately ruin of a mansion, a fortress, and a prison.   It was built in the middle of the fifteenth century by Ralph, Lord Cromwell, treasurer to Henry VI.   The chief features of the manor are a large groined vaulted undercroft or crypt, over which is a fine hall with an exquisite oriel window, a beautiful porch, and a well-preserved quatrefoil battlement.   Mary Queen of Scots was imprisoned here for nearly sixteen years.   The building suffered much during the siege at the time of the civil wars, and some of the cannon balls are preserved in the farmhouse near.

Chatsworth, "The Palace of the Peak," is a noted mansion, built in the sixteenth century by Bess of Hardwick, and pulled down to make way for the present building commenced in 1687 by William, the fourth

Haddon Hall

The Banqueting Hall, Haddon Hall

Earl and first Duke of Devonshire. It contains many art treasures, including some beautifully carved wood-work by Grinling Gibbons and a local artist Watson. The State apartments are on a grand scale, and have many times been occupied by royalty.   The large con-servatory, built by Sir Joseph Paxton, served him as a model for the 1851 Exhibition, now the Crystal Palace,

Wingfield Manor

and the gardens also contain the fine Emperor Fountain, which throws water to the height of 260 feet.

Of mansions of the time of Elizabeth and James I we have Hardwick Hall, Barlborough, Tissington, Sudbury, and numerous smaller ones. Dr Cox says there is no county in all England that has so many remnants left of the smaller halls and manor houses of the sixteenth and seventeenth centuries. Eyam, Swarkestone, Beeley,

Fenny Bentley, Bradbourne, Bradshawe, North Lees, Offerton, Riber, Youlgreave, Hartington, Highlow, and Snitterton are but a few of them.

"Hardwick Hall, more glass than wall," is a stately symmetrical building, the windows of which are overdone for the sake of effect. Some are merely shams. There are fine plastered rooms, and the original entrance gate-

Chatsworth

way and garden walls built by the renowned "Bess of Hardwick" still remain, whilst the parapet and flower beds have the initials E.S. and a coronet. Near by are the ruins of the older manor house.

Bolsover is interesting because it is built on the site of the old Norman keep, and is partly composed of the old materials. It has thick walls, quaint vaulted chambers, and fine chimney-pieces. Outside are the ruins of a later building, the riding school of the Duke of Newcastle. The four watch-towers are landmarks for miles around.

Derbyshire has no late seventeenth century work of the time of Inigo Jones and Sir Christopher Wren, but of the succeeding period when classic architecture was well established, we have an example in Kedleston, the

The Hall, Eyam

seat of Lord Scarsdale, the father of Lord Curzon of Kedleston.

Derby has a good example of a Jacobean house, which is now converted into a café.

The beautiful garden at Melbourne, laid out in 1720, is in the Dutch style, and contains stately vistas with fountains and statues at central view-points, and a dense yew-hedge walk.

The Presence Chamber, Hardwick Hall

26. Communications, Past and Present. Roads, Canals, Railways.

In these days of rapid locomotion, it is hard to realise the difficulty there was in past time for people who wished to travel from one part of the country to another. At one time the only means of communication was by British track-ways or forest paths which traversed parts of the country. One of these ways passed through Derby, and the river Derwent was crossed by a ford, probably near what is now known as the Holmes, which was only passable when the water was low. The Romans during their occupation of Britain made several roads through Derbyshire which are briefly described in the chapter on antiquities. After the withdrawal of the Roman legions, the existing roads were allowed to fall into decay, and new roads were not constructed. For many centuries the people journeyed by rude paths, only capable of being passed on foot or at best by horses. The Roman roads were generally made in a direct line across the country, over hills, down into the valleys, and up the other side irrespective of the gradient. In later roads the same method was followed, no doubt because they were used at first by pedestrians and riders on horseback when there were few or no wheeled conveyances. An instance of such roads is that from Chesterfield to Matlock, which was made down the steep descent of Matlock Bank to the bridge over the Derwent at Matlock Bridge. Later a more suitable course was chosen between town

Derby from the Derwent

and town, and hills and streams were avoided as much as possible.

In the eighteenth century turnpike roads were made. Towards the end of that century when coaches came into use for travelling, the roads were much improved.

The earliest turnpike Act that had reference to Derbyshire was for repairing and improving the road from the Trent at Shardlow through Derby and Hognaston to Brassington. The reason alleged for this first turnpike road in Derbyshire terminating at a small place such as Brassington was that the "traveller towards the north, having by means of this improved road been helped over the low and deep lands of the county and landed on the rocky districts might find his way therein, without further assistance, to Buxton, Tideswell, Castleton, Stoney-Middleton, Ashford, Bakewell, Winster, Matlock, Wirksworth, Hartington, Longnor, etc." The latter part of this road, which was hilly, was afterwards abandoned except by the natives, and travellers preferred to journey by Ashbourne to Newhaven, although it was a longer route.

In 1814, less than a century ago, the road from Derby to Matlock passed through Kedleston, Wirksworth, and Cromford. Before 1818 there was no direct public road from Derby to Cromford. On the western side of the Derwent was a private road belonging to the Strutts, the Harts, and Richard Arkwright. It was made into a public road in 1818.

Other old roads in the county were (1) the Nottingham and Newhaven turnpike which entered Derbyshire near Alfreton and passed through Wessington, Matlock, Snitterton, Wensley to Pike Hall, where it crossed the old Roman road and ended at Newhaven; (2) the road from Matlock to Ashbourne, which passed through Hopton, Cromford, and Kniveton; (3) that from Matlock to Alfreton and Nottingham, which passed through Cromford, Lea Bridge, Holloway, Crich, and South Wingfield;

Cromford Bridge

(4) that from Matlock to Buxton, through Darley, Rowsley, Bakewell, and Ashford, as well as by Newhaven.

The growth of trade caused an increase of traffic. The main roads from Manchester, London, Leeds, and Birmingham passed through Derby, so that the county town became one of the changing stations for the wagons and coaches on these roads. In 1735 Derby was in communication with London by coach, which ran once

a week. In 1790 the coach (then a through coach from
Manchester) left Derby daily about three in the afternoon
and arrived in London about ten o'clock the next morn-
ing. In the year 1828 there were at least seven coaches
each day from Derby to London, and from Derby to
Manchester, besides others to Nottingham, Birmingham,
Leeds, Newcastle-under-Lyme, and other towns.

Bridges are a necessary accompaniment of roads, and
often were built by a ford where the water was shallow.
The old county bridges, notably those at Cromford,
Matlock, Bakewell, and Rowsley were first erected for
pack-horse and pedestrian traffic only, and when wheeled
vehicles came into use they had to be widened. Proof of
the widening may be seen in the different style of architecture on opposite sides of several of the bridges.

Before the construction of railways, the only means
of conveying goods and minerals other than by road was
by canal. Great Britain was the last nation in Europe
to construct canals as a means of inland navigation.
Various expedients were adopted for removing obstruc-
tions in the rivers with a view of facilitating commerce,
but it was not until the year 1755 that the construction
of canals entered the system of British economy. A large
amount of capital was provided for the making of canals,
and risked in undertakings that were likely to prove
advantageous, and in a short time the country was inter-
sected by canals.

In 1719 an attempt was made to render the Derwent
from Derby to the Trent navigable, but in 1794 the
Derby canal from Derby to the Trent and Mersey canal

was navigable, and, according to Simpson, gypsum, stone, and other materials were imported to, and coal and cheese were exported from, Derby.

The Cromford canal, which was completed before the year 1794, began at Cromford, ran for some distance on the west side of the Derwent, and then from Lea followed the east bank to Ambergate, Whatstandwell,

Road from Castleton to Buxton and Chapel-en-le-Frith

and Bull Bridge, and joined the Erewash canal at Langley Bridge. Five miles to the north of Eastwood, a tramway worked by horses carried coals and cotton from the Pinxton wharf of the Cromford canal up to Mansfield, and brought back stone, lime, and corn to the canal. Cromford wharf was a busy centre for the import and export of produce. Matlock Bath and Cromford obtained their supplies of coal from there.

The whole of the stone quarried from Stancliffe for the building of St George's Hall, Liverpool, was shipped to its destination by the Cromford canal.

Nottingham, Derby, and Leicester were important centres of industry before the introduction of railways, holding constant communication with London and Birmingham and with each other. But the only modes of conveyance were by canal, by fly wagon, and by coach, and the charges made were proportional to the speed. Wool required two days to travel the 15 miles between Leicester and Market Harborough. Only three coaches ran daily each way between Leicester to Nottingham besides the through coaches from distant places. Many of the fly wagons were long stagers and were of little benefit to the intermediate towns.

The charge for conveying haberdashery from London to Leicester was £2. 15s. 0d. a ton by canal, 5s. a cwt. or £5 a ton by wagon, and 1d. a pound or over £9 a ton by coach.

One of the earliest railways constructed in England was the Cromford and High Peak Railway. It was opened in the year 1830, and was 33 miles in length. It began at the High Peak Junction near Cromford by the Cromford canal and ended at Whaley Bridge at the Peak Forest canal. Its course was over the hills and not along the valleys. Parts of the line were along inclined planes and the longest stretch of level ground was 12 miles. This railway not only carried goods from one canal to the other over the Pennine Chain, but also opened out a large mineral trade in the hilly district over which it passed at a later date. The line was taken over by the London and North-Western Railway, and is now connected with the Midland near Cromford and the London and North-Western Railway at Parsley Hay and Buxton, the portion between Buxton and Whaley Bridge having been dismantled.

The rise of the Midland Railway, which has for many years been associated with the county town of Derby, is an interesting one. The Leicester and Swannington line was opened in 1832. The coal raised in the Nottinghamshire and in the Leicestershire coalfields were distributed mainly in the respective areas by water navigation. The attempt of the Erewash Valley coal-owners to obtain a market for their coal in Leicester resulted in the construction of the Midland Counties Railway, which connected Nottingham, Derby, and Leicester. It was opened in 1839. In the same year the line from Birmingham to Derby was opened, and a year later the North Midland line from Leeds to Derby. In the year 1844 an amalgamation of the three railways took place and formed the nucleus of the Midland Railway. This amalgamation made Derby an important railway centre, and since that time the Midland Railway has been greatly extended into a complete system in England, Scotland, and Ireland. The Midland is the only railway whose main line passes through the county, but there are branches of several other important lines which serve the needs of Derbyshire. The London and North-Western Railway, in addition to the High Peak line, runs from Manchester to Buxton, thence to Ashbourne and Uttoxeter, and afterwards joins its main line through Rugby to London.

Midland Railway Tunnel in Mountain Limestone near Monsal Dale

The North Staffordshire Railway has a line from Derby to Uttoxeter and Ashbourne, and the London and North-Western Railway has running powers over this line from Ashbourne to Uttoxeter.

The Great Northern has a branch line from Nottingham to Burton *viâ* Egginton, and during the last few years the Great Central has constructed a line through Chesterfield to Manchester.

It will thus be seen that Derbyshire is very well supplied with railways, notwithstanding the hilly nature of the country.

The introduction of railways was vigorously opposed by the canal owners, who naturally objected to being deprived of a monopoly in which they had invested large sums of money, and some of the canals had to be purchased by the railways and kept up by them. This resulted in railways spending large sums of money on canals which they would rather not have possessed. What future there may be for canals is a matter of speculation. The use of horse-power renders them slow in transit. It is possible, however, that as the motor car on roads is revolutionising traffic and competing with railways to some degree, that a similar motive power applied to barges might make canals more useful. But against this, there is the fact that they cannot be used in hilly districts because of the numerous locks and consequent delay in traffic and the large quantity of water which would be required and which would be impossible to obtain. The transhipment from canal to railway is also a factor which increases the expense of conveyance.

27.   Administration and Divisions— Ancient and Modern.

The government of our English counties is the result of a blending and alteration of several systems. The word Sheriff is a survival both of the Saxon and Norman rule. In Saxon England the county or shire was presided over by the Shire-reeve, who was elected by the people. In Norman England the Count, who represented the Sovereign, was head of the county. In time he deputed his functions to some local lord, who became known as the *Vicecomes* to the Normans, but by the more English name of Shire-reeve or Sheriff to the people, who preferred the old Saxon to the new Latin word. This is why the Saxon Sheriff, originally elected by the people of the county, has for many centuries been appointed by the Sovereign of England.

Derbyshire formed part of the Saxon kingdom of Mercia. The eastern portion of that kingdom was conquered by the Danes, and Derby became one of the five Danish burghs and was therefore associated with the other four burghs of Lincoln, Leicester, Stamford, and Nottingham. During the Norman rule the shires of Derby and Nottingham were grouped together for administrative purposes. For this reason, up to the reign of Henry III, the assizes of the two counties were held at Nottingham, and there was only one county gaol, that at Nottingham, for both shires. From that time up to 1566, the assizes were held alternately at Derby and

Nottingham. In the year 1566, up to which time there had only been one Sheriff for both counties, an Act of Parliament was passed giving a separate Sheriff to each of the counties of Derby and Nottingham.

The office of Lord-Lieutenant was not created until the year 1554. Hallam remarks that "the power of calling to arms and mustering the population of each county given in earlier times to the Sheriff or Justice of the Peace or to Special Commissioners of Array began to be entrusted to the Lord-Lieutenant." His "office gave him the command of the Militia, and rendered him the chief Vice-regent of his Sovereign responsible for the maintenance of public order." The Lord-Lieutenant was at first an extraordinary magistrate, constituted only in time of difficulty and danger, but the office soon became a permanent institution.

Derbyshire is at the present time divided into six hundreds or wapentakes, viz. the High Peak, Wirksworth or the Low Peak, Appletree, Scarsdale, and the joint hundreds of Morleston and Litchurch, and of Repton and Gresley. Modern historians consider that the hundreds of the Domesday Book cannot be correlated with those of the present day, because in ancient records the terms for these divisions, such as wapentake, hundred, liberty or soke appear to have been used somewhat indiscriminately.

We will now consider the present mode of county government. The chief officers in the county are the Lord-Lieutenant and the Sheriff, who are appointed by the King, the latter annually and the former generally for life. The Lord-Lieutenant is in most cases a nobleman or large landowner. In Derbyshire, the Lord-Lieutenancy has been held by the nine Dukes of Devonshire, and the present Duke is the twenty-second person who has held that office.

The County Council now conducts the main business of the county, and meets in the County Council buildings in St Mary's Gate, Derby, where the offices and Council Chamber are situate. This County Council consists of 21 Aldermen and 63 Councillors, the latter of whom are elected, nine from each of the Parliamentary Divisions of the County. The County Council, amongst other duties, keeps the county roads and bridges in repair, appoints the police, manages the County Lunatic Asylum, and looks after the health of the people; and generally carries into effect the laws passed by Parliament.

The County Council represents the central form of County Government which was started in 1888, but another Act was passed in 1894 for Local Government in the towns and parishes. In the large parishes, the chief authorities are now the District Councils, of which there are 26 in Derbyshire, while the smaller parishes have their Parish Council or Parish meetings.

In Derbyshire there are four boroughs which have been incorporated by Royal Charter. They are each governed by a Mayor and Town Council. Derby is the most ancient borough in the county, has the powers of a County Council and is called the County Borough of Derby. The governing body consists of 16 Aldermen and 48 Councillors. The Borough of Chesterfield, which

was incorporated in 1204, has six Aldermen and 18 Councillors; that of Glossop the same number of each as
Chesterfield, and that of Ilkeston five Aldermen and
14 Councillors. The boroughs have greater powers of
government than the parishes.

Derbyshire is divided into nine Poor Law Unions,
each of which is under a Board of Guardians, whose
duty is to manage the workhouses and to appoint and
control various officers to carry on the work of relieving
the poor.

For purposes of justice, the Administrative County
has one court of Quarter Sessions, which meets at Derby,
and is divided into 15 Petty Sessional divisions, each
having Magistrates or Justices of the Peace, whose duty
it is to try cases and administer justice. The County
Borough of Derby and the Municipal Boroughs of Chesterfield and Glossop have separate Commissions of the
Peace.

The number of civil parishes in Derbyshire is 314.
There are 255 ecclesiastical parishes within the ancient
county of Derby, and of these 249 are in the Diocese of
Southwell, three in that of Peterborough, and three in
that of Lichfield. Formerly Derbyshire was in the
Diocese of Lichfield, but now Derbyshire and Nottinghamshire form the Diocese of Southwell.

The control of public elementary and secondary
education in the county under the Education Act, 1902,
is directed by five Education Committees, elected respectively by the County Council, the Council of the County
Borough of Derby, and the Councils of Chesterfield,

Glossop, and Ilkeston.   Since 1902, further powers have been given to, and duties placed upon, the Education authorities, including amongst the former the feeding of necessitous children in elementary schools, and amongst the latter the medical inspection of children in elementary schools.

Derbyshire is represented in the House of Commons by nine Members of Parliament.   The County Borough of Derby elects two Members, and the remainder of the county, which for Parliamentary purposes is divided into seven divisions, sends one Member from each division.

28.   The Roll of Honour of the County.

There is an old adage which says :—

> " Derbyshire born and Derbyshire bred,
> Strong in the arm and weak in the head,"

or, as some say it should be, "wak" meaning "awake in the head."   Be this as it may, Derbyshire is not without her illustrious men and women in every walk of life.

The first Earl of Derby was the powerful Norman Baron, Robert de Ferrers, who owned so much land in this county.   The title was forfeited in the reign of Henry III, but Henry VII revived it, and conferred it upon Thomas, Lord Stanley, who had crowned him King.

Amongst more recent statesmen we have Lord Melbourne, the first of Queen Victoria's Prime Ministers.

His name was taken from the little Derbyshire village where his ancestors, the Coke family, resided, and it has been adopted for that of one of the great capitals of our Australian colonies. Lord Chief Justice Denman, born in 1779, owned a farm at Stoney Middleton, and his grandfather was a doctor at Bakewell. Lord Denman was a reformer of abuses, and truly an upright judge.

The Dukes of Devonshire have resided at Chatsworth, and the Cavendish family have been represented in the House of Commons almost continuously for over three centuries. The noted " Bess of Hardwick " built three of the finest seats in the county, viz. Chatsworth, Hardwick, and Oldcotes, but Chatsworth has been rebuilt and Oldcotes has disappeared. Horace Walpole records a tradition that she would die if she ceased building, and that shortly before her death, the work at Bolsover Castle was stopped by the frost. But alas for the tradition, Bolsover was built by her son ! She was married four times, and remained a widow for the fourth time for 17 years. Her second husband was William Cavendish, and her fourth husband George, Earl of Shrewsbury, the custodian of the unhappy Mary Queen of Scots.

Amongst soldiers there was Sir John Gell of Hopton, who made himself notorious over the collection of ship-money, though better known, perhaps, was his kinsman of more modern times, Sir William Gell, the archaeologist and geographer—" topographic Gell " as Byron terms him—who wrote on Pompeii, Rome, and Greece generally and died in 1836. Derbyshire can proudly point to Lieutenant-General Sir James Outram of Butterley

Hall, "the Bayard of India," as one of her most noble and famous sons, a tower of strength during the Indian Mutiny, and in turn both besieger and besieged in Lucknow. He died in 1863 and was buried in Westminster Abbey.

Notable amongst Derbyshire divines, were Bishop Pursglove, who was born and bred in Tideswell, and died in 1579, having built two grammar schools and a hospital ; William Bagshawe, "the Apostle of the Peak," an eminent Nonconformist and author, who was born at Litton in 1628; and Mompesson the vicar of Eyam, whose courageous unselfishness at the time of the Plague did so much to restrict its spread.

Of men of letters the most salient name is Richardson, though less by virtue of peculiar genius than by the fact of his having struck a new vein in fiction which appealed to the readers of that day with a success which nowadays seems astounding. Born in 1689, the son of a joiner, and with no special education, Samuel Richardson was still more remarkable in that his first novel *Pamela* was published when he was over fifty. It was followed by *Clarissa* and *Sir Charles Grandison*, but all three have passed long since into practical oblivion.

William Howitt, born at Heanor in 1792, and his wife, Mary Howitt, were keen lovers of nature, and wrote many poems as well as tales and travels which are both interesting and instructive, amongst the latter an account of the Australian goldfields in early days.

Newton, "the Minstrel of the Peak," who was born near Tideswell and lived at Eyam, is better known within

Samuel Richardson

the county than to the outer world; Antony Bradshawe of Farley's Hall, the author of a quaint and interesting poem on Duffield and the customs of the peak, was a barrister and a historian. His memorial, self-erected fourteen years before his death in 1614 in Duffield church, is inscribed to himself and his two wives and twenty children.

William Hutton, the well-known historian of our county, was born in Full Street, Derby, in 1723. Besides his history, he wrote a large number of poems, and an interesting autobiography, which reveals his hard fight with poverty, and the life of the lower classes of his time.

The famous philosopher of the nineteenth century, Herbert Spencer, was born on April 27th, 1820, at Exeter Row, and lived later in Wilmot Street, Derby. He was distinguished for having applied the principles of evolution to the social and moral development of mankind; and the ten volumes dealing with his stupendous system of synthetic philosophy mark thirty-six years of unflagging industry against adverse circumstances and continuous bad health. His best known work on *Education, Intellectual, Moral, and Physical*, has been translated into nearly every known language including Chinese and Japanese, and a copy is given to every public teacher in France. He died December, 1903, and left an interesting biography which was published after his death.

Erasmus Darwin, poet, evolutionist, and physician, and grandfather of the illustrious Charles Darwin, was the author of *Loves of the Plants*, and was a resident in Derby for nearly twenty years, though not a native.

Derbyshire is particularly noteworthy for its scientific men. John Flamsteed or Flamstead, the first Astronomer Royal, was born of Derby parents, who went to Denby, five miles away, to escape the plague in 1646, the year after

John Flamsteed

his birth. He went to Greenwich the year the Observatory was built (1676) and formed the first proper catalogue of the fixed stars. He was a friend of Isaac Newton, and his great work was the *Historia Caelestis Britannica*.

The Derbyshire toadstones furnished John Whitehurst with material for his speculations, which were published in 1778. He was the first to believe that they were as truly igneous rocks as those of Vesuvius or Etna. The family of Whitehurst was well known as clockmakers in Derby.

Amongst artists, Joseph Wright, always known as "Wright of Derby," who was born in 1734, bears a deserved reputation. He was a wonderful painter of artificial light, of fire, moonlight, volcanoes, and other studies in chiaroscuro. Some of his finest works are portraits and historical subjects, several of which find a permanent home in the Derby Art Gallery.

Sir Francis Chantrey, the matchless English sculptor, was born at Norton in 1781. His beautiful monument of "The Sleeping Children" at Lichfield is well known.

Brindley, the famous engineer and constructor of canals, was born at Thornsett in 1716. He could hardly read or write, but took Nature as his book, and his inborn mechanical genius solved most of his problems.

The founders of the great cotton industry, "Arkwright, Strutt, and Need," built their first mills in Derbyshire at Cromford and Milford, and later acquired residences in the neighbourhood. Jedediah Strutt invented the ribbed stocking machine in 1758. Joseph Strutt presented the Derby Arboretum to the town, and many other proofs of this family's munificence have been given to Derbyshire.

We must also record Michael Thomas Bass as a Derby benefactor. Though not a Derbyshire man, he represented the borough in Parliament for thirty years. He was simple-minded and unobtrusive, but princely in his gifts, among which were the Free Library, Museum, Art Gallery, and Swimming Baths.

Other notabilities who resided in Derby during the

Sir R. Arkwright

eighteenth century were Duesbury, the Staffordshire potter, who established the Derby China works in 1755 in connection with Heath, of the Cockpit Hill works; and John Lombe, who started the silk-throwing industry in Derby.

Finally, a heroine of whom Derbyshire will ever be justly proud is the venerable lady, still with us, Florence Nightingale of Lea Hurst, near Cromford. She was wealthy and accomplished, but she loved nursing and studied it as a science, so when the call for help came from our suffering and neglected soldiers in the Crimea, she was willing and prepared to obey the call. She revolutionised nursing, and when a grateful nation subscribed £50,000 as a recognition of her services, she devoted it all to founding the Nightingale Home for Trained Nurses.

Hardwick Hall

29. THE CHIEF TOWNS AND VILLAGES OF DERBYSHIRE.

(The figures in brackets after each name give the population from the 1901 census returns, and those at the end of each section are references to the pages in the text.)

Alfreton (17,505) is a market town, pleasantly situated on the brow of a hill, and connected by the Midland Railway with Mansfield. The inhabitants are chiefly employed in the collieries and iron-works of the neighbourhood. (pp. 71, 132.)

Allestree (589). A parish two miles north of Derby. (pp. 102, 106.)

Ambergate and **Heage** (2490). A large parish. Ambergate station is a junction on the Derby, Chesterfield and Sheffield and on the Derby, Matlock and Manchester branches of the Midland Railway. (pp. 25, 76, 134.)

Ashbourne (4039) (mentioned in Domesday Book as Esseburn), on the slope of a hill, is a market and union town. It is 13 miles from Derby by road, and four miles from Dovedale, and is the terminus of the branch of the North Staffordshire Railway from Rocester, and also of the branch line of the London and North-Western Railway from Buxton. It is an interesting old town, with a fine parish church, noted for its spire, which is called the

"Pride of the Peak," and a grammar school founded in 1585. Its trade depends on the farmers and the numerous fairs and markets. The Royalist troops suffered defeat here in 1644, and Ashbourne Hall, now converted into a hotel, was occupied by the Pretender on his march to Derby, and also on his retreat from that town. On Shrove Tuesday, the old custom of playing football in the streets is still kept up. (pp. 34, 35, 74, 76, 81, 83, 89, 103, 104, 106, 107, 109, 111, 131, 132, 136, 138.)

Ashford (684), often called Ashford-in-the-Water, is a parish about a mile and a half from Bakewell on the river Wye, which here flows between lofty hills. Up to a few years ago, it was celebrated for its marble quarries, but the industry has now become extinct. Rottenstone was obtained from the neighbourhood. (pp. 121, 131, 132.)

Ashover (2426), seven miles south-west of Chesterfield, is a township on the Amber, which here flows through the Mountain Limestone. Extensive limestone quarries are worked, but the many old lead-mines are disused. (p. 102.)

Aston-upon-Trent (537). A parish and village six miles south-east of Derby. All Saints' church contains a Saxon Cross built into the wall and some portions of late Norman work. (p. 102.)

Ault Hucknall (1582). A village and parish seven miles from Chesterfield. The church of St John the Baptist is Norman and Early English. Hardwick Hall, a seat of the Duke of Devonshire, was erected in the reign of Queen Elizabeth. (p. 98.)

Bakewell (2850) is an interesting market and union town beautifully situated on the Wye, which is spanned by a stone bridge of six arches. It has a station on the Midland Railway, and is noted for its fine church and old cross. Several chert quarries are worked, and the rock sent to the potteries. Haddon

Hall and Chatsworth are in the immediate neighbourhood.   (pp. 71, 74, 83, 89, 102, 107, 111, 121, 131, 132, 133, 144.)

Barlborough (2056).   A village and parish eight miles from Chesterfield.   St James's church contains four Norman arches.   The chancel is Early English.   (p. 125.)

Beighton (3371).   A village and parish in the north-eastern division of the county.   The chapel of St Thomas has a western tower in the Norman style built of part of the ruins of the abbey in 1660.

Bridge over the Wye, Bakewell

Belper (10,934), a market and union town on the Derwent about eight miles due north of Derby, is a straggling modern town, with a station on the Midland Railway.   The chief industries are cotton-spinning, and the manufacture of hosiery, stoves, and grates.   (pp. 59, 68.)

Blackwell (4144).   A parish about three miles from Alfreton with extensive collieries which employ a large number of people.   (p. 107.)

Bolsover (6944) is a large village six miles east of Chester-field on the Chesterfield and Mansfield Railway. Bolsover Castle is in the neighbourhood. (pp. 76, 82, 103, 117, 126, 144.)

Bonsall (1360). A town and parish about one and a half miles from Cromford Station, situate about 700 feet above sea level. Some of the inhabitants are employed in limestone and dolerite quarries, and in paint and colour works. The Via Gellia, a beautifully wooded and narrow valley in the limestone, is partly in this parish. (pp. 68, 104.)

Bradbourne (132). A township, parish, and village six miles from Wirksworth. All Saints' church contains remains of Saxon and Norman work. (pp. 102, 107, 126.)

Bradwell (1033). A township and old town two miles from Hope station, situate about 600 feet above sea level and nearly surrounded by limestone hills. Here are large limestone quarries and the Bagshaw Cavern. (pp. 14, 46, 94, 95.)

Brampton (2185). A township, village, and parish adjoining Chesterfield on the west. The church contains a Norman doorway.

Brassington (651). A township, village, and parish four miles from Wirksworth, 800 feet above sea level. The church possesses extensive Norman remains. In this parish are several brickworks and limestone quarries, Harboro' and Rainster Rocks, and the Hoe Grange Quarry, in which a cavern with Pleistocene remains was discovered. (p. 131.)

Breadsall (515). A village and parish two and a half miles from Derby with a station on the Great Northern Rail-way. The ancient church contains many interesting remains, including a carved alabaster "Pieta" and some chained books. A priory was founded here in the reign of Henry III, and the estate was at one time owned by Dr Erasmus Darwin. (pp. 102, 103.)

Brough (66). Brough and Stratton form a township near Hope.   At the confluence of the rivers Noe and Bradwell are the remains of a Roman Camp.   (pp. 60, 93, 94, 95.)

Burbage (1503).   On the Wye, one mile from Buxton. It was formed into a civil parish in 1894.   Poole's Hole, a cavern in the limestone, is in the parish.   The large quarries belonging to the Buxton Lime Firms Co. Ltd. employ several hundred men and have a branch line from the L. & N. W. Railway.   (p. 17.)

The Crescent, Buxton

Buxton (6373), close to the eastern border of the county and the source of the Wye and called Aquae by the Romans, is one of the important inland watering-places of England.   It is situate at a height of 1000 feet above sea level, the portion called Higher Buxton being about 100 feet higher.   It has two railway stations, the Midland and the London and North-Western Railway.   The town is placed in the deep valley of the Wye, and is almost surrounded by hills.   The district is wild and bleak, but beautiful river scenery is afforded by the Goyt valley and by the deep gorges of the Wye in the direction of Miller's Dale. The chief interest is the spring of tepid water which issues from the ground at a temperature of 82 degrees and is largely charged with nitrogen. It is used for bathing and medicinal purposes, and there are a number of various kinds of baths in connection with it. Poole's Hole, a cavern in the limestone, is close to the town; and there are limestone and sandstone quarries in the neighbourhood. (pp. 8, 14, 15, 17, 44, 46, 50, 56, 57, 59, 76, 77, 86, 93, 95, 131, 132, 136.)

Castleton (547) is a village about two miles from Hope Station. It is celebrated for its ancient Peak Castle, the ruins of which remain, the Peak Cavern, Speedwell Cavern, and "Blue John" mine, The Winnats, and Mam Tor or the Shivering Mountain. It is visited by a large number of people. (pp. 12, 40, 42, 43, 44, 46, 48, 76, 82, 93, 94, 116, 131.)

Chapel-en-le-Frith (4626), a small market town, is about six miles north of Buxton on the Midland and London and North-Western Railways. (pp. 43, 86.)

Charlesworth (1967). A township and parish about two miles from Glossop and near the borders of Cheshire. The chief trades are cotton spinning, rope making, and cotton-band making.

Chelmorton (287). A parish five miles from Buxton and eight from Bakewell. St John the Baptist's church contains traces of Saxon and Norman work. It is the second highest village in England (1218 feet above sea level). (pp. 107, 109.)

Chesterfield (27,185) is a market town, and the largest of the three Derbyshire boroughs. It is noted for its church with a twisted spire. The Stephenson Memorial Hall was built in memory of George Stephenson, the railway engineer, who lived near Chesterfield. (pp. 21, 84, 86, 88, 95, 104, 105, 107, 111, 115, 116, 130, 138, 142, 143.)

Castleton

Chinley (1223). Chinley, Bugsworth, and Brownside form a township in the parish of Glossop, and together contained 1223 inhabitants in 1901. Chinley is an important junction on the Midland Railway, which here passes over two fine viaducts. (p. 94.)

Church Gresley (8618). A parish and township six miles from Ashby-de-la-Zouch. A priory of Austin Canons was founded here in 1135–40 by William de Gresley. Part of the aisle, arcade and the tower of the existing church are all that remain, though many fragments of Norman work have been found in the churchyard. There are important coal-works, potteries, fire-brick and encaustic tile works.

Clay Cross (8358). A town five miles south of Chesterfield with a station on the Midland Railway. The Urban District Council includes Clay Lane (7701), and Egstow (657). About 3000 hands are employed by the Clay Cross Company (Iron).

Clowne (3896). A village and parish with a station on a branch of the Midland Railway and also on the Great Central Railway about eight miles from Chesterfield.

Codnor and **Loscoe** (3831). These are hamlets forming a civil parish on the Heanor and Ripley branch of the Midland Railway. A great number of men are employed in the extensive collieries of the Butterley Company.

Crich (3063). An old town 600 feet above sea level on an inlier of mountain limestone, and one mile from Whatstandwell station on the Midland Railway. The church of St Michael is partly Norman, but mixed in style   Crich Stand, 950 feet above sea level, is a noted landmark—a circular tower built in 1851; it is now closed to the public, as it is unsafe and is being allowed to fall into decay. An extensive landslip occurred here in 1882. A considerable number of people are employed in the limestone and gritstone quarries.   A large covered service reservoir is being constructed by the Derwent Valley Water Board from which pipes are being laid to Nottingham, Derby, and Leicester.   (pp. 37, 104, 109, 132.)

Cromford (1080).   A town and parish with a station on the Manchester line of the Midland Railway.   Sir Richard Arkwright established the first cotton factory here in 1771.   (pp. 26, 28, 30, 33, 68, 71, 131, 132, 133, 134, 135, 136, 149, 151.)

Darley Abbey (915).   A parish and village one and a half miles north of Derby and on the river Derwent.   The ancient abbey of St Mary was founded before 1112, by Hugh, dean of Derby.   (pp. 84, 113, 115.)

Darley Dale (2756) (or **North Darley**) has a station on the Midland Railway about four miles from Matlock Bath.   The church dates from Norman times and contains a Norman font. Here are the Whitworth Institute and Hospital, and sandstone quarries are worked.   (pp. 16, 17, 53, 77, 107, 109, 132.)

Denby (1731).   A parish and village eight miles north of Derby with large collieries, iron furnaces, and pottery and brick tile-works.   (p. 148.)

Derby (105,912), on the Derwent, 127 miles N.N.W. of London, is the chief town of Derbyshire, a County Borough and Parliamentary Borough, and the headquarters of the Midland Railway.   Owing to its position as an important railway centre near the coalfields of the Midlands, it is essentially a manufacturing town, and many and varied industries are carried on in it.   It has a number of churches, the most interesting being All Saints', with a beautiful tower.   The Free Library, Museum and Art Gallery, and Free Swimming Baths were built and presented to the town by the late Michael Thomas Bass, M.P., and the Arboretum by Joseph Strutt.   St Helen's House has for many years been used for Derby School, which is one of great antiquity. (pp. 3, 21, 25, 28, 57, 59, 61, 62, 63, 67, 68, 69, 74, 83, 84, 85, 86, 87, 88, 107, 111, 113, 114, 115, 118, 127, 130, 131, 132, 135, 136, 138, 139, 142, 143, 147, 148, 149, 150.)

Dethick and **Holloway** (1311). This is a parish two and a half miles from Matlock. The church dates from 1220. At Holloway is situated Lea Hurst which was for many years the home of Florence Nightingale.

Dore (1305). Dore is a village which, since 1844, has formed a separate parish with Totley. It has a station on the Midland Railway five miles south-west of Sheffield. The Dore and Chinley branch of the Midland Railway, 20 miles in length, passes through five and a half miles of tunnels and the remainder through Derwent Dale, Hope Dale and Edale. (p. 94.)

Dronfield (3809) is a manufacturing town six miles from Chesterfield on the road between Chesterfield and Sheffield. (pp. 84, 106, 109.)

Duffield (1959) is situated on the Derwent and Ecclesbourne brook about four miles north of Derby. After the Conquest it became the seat of the de Ferrers, Earls of Derby. The foundations of Duffield Castle on Castle Hill have been laid bare, and the site has been presented to the parish. (pp. 14, 68, 83, 116, 118, 147.)

Eckington (12,895). A township and parish with a station on the Midland Railway and is one mile from Renishaw station on the Great Central Railway. The church dates from the time of Stephen and has considerable Norman remains. (p. 103.)

Eyam (269). A township, village and parish five miles east of Tideswell and six miles north of Bakewell. The village is picturesque and interesting, and the churchyard contains a Saxon Cross and the tomb of Catherine Mompesson, wife of the Derbyshire hero of the Plague. (pp. 44, 85, 107, 125, 145.)

Glossop or **Glossop Dale** (21,526), a municipal borough and market town, is on the borders of Cheshire. It is the chief seat of the cotton manufacture in Derbyshire, having extensive factories besides woollen and paper mills, while calico printing is also carried on in the neighbourhood. (pp. 68, 94, 142, 143.)

Hartington (3306) is a parish consisting of the four townships called Middle, Upper, Nether, and Town Quarters. Hartington Town Quarter has a station one and a half miles away from the village on the Buxton and Ashbourne branch of the L. & N. W. Railway. In the Upper Quarter is Axe Edge, 1756 feet above sea level, from which the rivers Dove, Wye, Dane, and Goyt have their sources. (pp. 126, 131.)

Hartshorne (1375). A village between Derby and Ashby-de-la-Zouch, two miles north-east from Woodville station. (pp. 68, 118.)

Hathersage (1135). A village with a station on the Dore and Chinley branch of the Midland Railway, ten miles from Sheffield. There are large gritstone quarries in the neighbourhood. Pins and steel wire are the chief manufactures. The place is situated in the midst of beautiful moorland and river scenery. (pp. 93, 104, 106.)

Hayfield (2614). A village about 600 feet above sea level. It is a convenient starting point for Kinder Scout, and possesses large calico-print works and paper and cotton mills. (p. 68.)

Heanor (12,418). A parish which includes Heanor, Langley Mill, Langley Marlpool, and Aldercar and is situated on the road from Derby to Mansfield. (pp. 17, 67, 145.)

Hope (382). A village with a station on the Dore and Chinley branch of the Midland Railway two miles from Castleton. (pp. 43, 94, 107, 111, 115.)

Ilkeston (25,384), on the Erewash, is a municipal borough, with various manufactories. (pp. 67, 68, 83, 103, 104, 109, 143, 144.)

Killamarsh (3644). Called in Domesday "Chinewald-marese" is a widely scattered parish on the borders of Yorkshire and has a station on the branch of the Midland Railway and the main line of the Great Central, and on the Lancashire, Derbyshire and East Coast Railways The church is partly Norman. There are collieries, a steel forge, and chemical works.

Long Eaton (13,045). On the Erewash, one mile from Trent station. There are extensive railway carriage works; and lace-making is the chief occupation. (pp. 67, 102.)

Longford (352). A village six miles from Ashbourne and six miles from Tutbury. The church contains Norman piers and a Norman font. (pp. 102, 107, 118.)

Marston Montgomery (341). A scattered village two miles from Rocester station on the North Stafford Railway. The chancel arch of the church is tenth or eleventh century work and there is a south door, a priest's door in the chancel, and a circular font all of Norman date. The churchyard contains a fine old yew tree and some curious old tombs. The register up to 1660 was common to this place and Cubley and is still kept in the last named parish. (p. 98.)

Matlock (5979) parish includes Matlock Town and Green, Matlock Bank, and Matlock Bridge. The station is Matlock on the Midland Railway. Matlock Bank, which is sheltered from the east winds, is noted for its hydropathic establishments. Riber Castle, built by the late Mr Smedley, is now used as a boarding school for boys. (pp. 16, 17, 28, 39, 41, 46, 48, 95, 130, 131, 132, 133.)

Matlock Bath (1819) is an inland watering-place in the deep dale or gorge of the Derwent. It is situated amongst charming

Matlock Bath

limestone scenery, and is of world-wide repute for its medicinal springs. Of these there are three, which issue from the limestone at a temperature of 68 degrees Fahr. There are also petrifying wells and several interesting caverns. (pp. 7, 8, 28, 37, 39, 47, 134.)

Melbourne (3580) is a small town with manufactures about eight miles south-east of Derby, noted for its Norman church and the gardens of Melbourne Hall, once the seat of Lord Melbourne, which were formed in 1720 in the Dutch style. The Castle was dismantled in the fifteenth century. (pp. 81, 83, 86, 100, 109, 111, 117, 127, 144.)

Mellor (1218). A parish two miles from Marple station on the Midland Railway. Wadding and surgical dressings are manufactured in the neighbourhood. (pp. 102, 107.)

Mickleover (2084). A parish three miles from Derby with a station on the Great Northern Railway. The County Lunatic Asylum for 776 patients is located here. (pp. 109, 118.)

Middleton-Stoney (478). A picturesque and hilly township in the parish of Hathersage, five miles from Bakewell. It has two springs, one with a temperature of 60 degrees and baths have been erected upon the site of a supposed Roman bath. It possesses manufactories for barytes, and has several lime kilns. (pp. 131, 144.)

Milford (1096). The English Sewing Cotton Co. Limited, have here a large factory for bleaching and dyeing, which was originally founded by Messrs Strutt about 1780. (pp. 68, 149.)

Newbold-cum-Dunston (5986), adjoins Chesterfield. The ancient Norman and Perpendicular church was nearly destroyed by a mob in the reign of William III, and was desecrated and used as a cow-house. There are potteries for manufacturing stone bottles and brown ware and a brick and tile manufactory.

New Mills (6253). A manufacturing town on the river Goyt. (pp. 8, 10, 68.)

Norton (11,875). A pleasant village two and a half miles from Beauchief. The church has a few late Norman remains. In the churchyard is the grave of Sir Francis Chantrey, R.A., the eminent sculptor. (pp. 71, 149.)

Ockbrook (2567). A parish one mile from Borrowash station on the Nottingham and Derby branch of the Midland Railway. The tower and font of the church are Norman, though the spire is later, and the finely carved oak screen is of the sixteenth century. (p. 103.)

Peak Forest (476). A small village five miles from Buxton, with a station on the Midland Railway nearly three miles away and one on the L. & N. W. Railway. It was originally a free chapelry in the King's forest and was extra-episcopal and extra-parochial and up to 1804 there was an average of 80 "Gretna" weddings annually. There are extensive lime works near the Midland station. (pp. 14, 41, 44, 76.)

Pinxton (2994). A parish four miles from Alfreton noted for its collieries. (p. 134.)

Pleasley (8448). A pleasant village nine miles from Chesterfield. The church contains a highly ornamented Norman arch. Pleasley Vale is the site of large cotton, silk, and merino spinning mills near picturesque limestone ravines.

Repton (1695) is celebrated for its old priory and its school, which was founded in 1556 by Sir John Port. It is about seven and a half miles S.S.W. of Derby. (pp. 61, 68, 99, 107, 111, 112, 118.)

Ripley (10,111) is a market and manufacturing town. The ironworks and collieries of the Butterley Co. are in the immediate neighbourhood. (pp. 17, 68.)

Rowsley (295). Has a station on the Midland Railway and is five and a half miles from Matlock. The district is noted for its manufacture of grindstones and for excellent building stone. (pp. 132, 133.)

Sandiacre (2954). A large village. The church of St Giles contains traces of Norman work. (pp. 35, 83, 102, 103, 104, 106, 107, 109.)

Sawley (1751), has a station one and a half miles south-east of the village and another (Sawley Junction) half a mile south-west on the Midland Railway. A church existed here previous to 822, and the north wall of All Saints' church contains herring-bone work and is supposed to be Saxon. (pp. 95, 98, 104, 107.)

Spondon (2544). A parish and township with a station on the main line of the Midland Railway, large colour-works, and a tar distillery. (pp. 107, 109.)

Stanley (1263). About a mile from the West Hallam station on the Great Northern Railway. The small church possesses a Norman south door. (p. 114.)

Staveley (11,420). A large village with two stations on the Midland Railway and two on the Great Central. The church contains some interesting tombs and incised slabs, and a fine font of twelfth century date. The land yields rich minerals and an abundant supply of coal, and there are also corn-mills, a brush manufactory, and one for spades and shovels.

Swadlincote (4017). Noted for the manufacture of earthenware and fire-bricks. It has a station on the Burton and Ashby branch of the Midland Railway and is six miles from Burton-on-Trent. (p. 68.)

Swarkestone (146), a parish and village three miles south-east of Derby. It is remarkable for its ancient bridge and raised causeway nearly a mile in length, the oldest remaining portions being of thirteenth century date. There once stood a midway chantry or chapel upon it. There are some slight Norman remains in the church and a probable Norman font. Near the site of the "Old Hall" is a balcony and enclosure supposed to have been used for bull-baiting. (pp. 87, 88, 125.)

Tideswell (1938) is a market town two and a half miles from Miller's Dale Station. It is noted for its fine church with an unusually large chancel. The Grammar School was founded in 1560 by Robert Pursglove, Bishop of Hull. (pp. 14, 39, 84, 104, 107, 111, 131, 145.)

Tissington (367). A parish and village with a station on the Buxton and Ashbourne section of the L. & N. W. Railway, is four miles from Ashbourne. The church dates from Norman times and the Hall is a fine Elizabethan building. Tissington is remarkable for its ancient custom of well-dressing. The five wells which supply the village with water are elaborately decorated with flowers on Holy Thursday and a special service is held in the church.

Wensley (or **South Darley**) (788). An ecclesiastical parish separated from North Darley or Darley Dale by the river Derwent. It contains the celebrated lead-mine called Mill Close. (p. 131.)

Whitwell (3380). An agricultural and mining village and parish in the extreme north-east of the county, with a station on the Mansfield and Worksop branch of the Midland Railway. The church dates from 1150 and contains a Norman font. At Steetley, a farm one and a half miles north of Whitwell, is the small Norman church of All Saints. Whitwell Wood extends over 440 acres. (pp. 8, 101, 109.)

Wilne and **Draycott** (1504). Wilne is a parish about one and a half miles south from Draycott station on the Midland Railway. (pp. 99, 104, 106, 109.)

Wingfield, North (2973). A township and parish four miles south of Chesterfield celebrated for its extensive coal, lime, and ironstone beds. There are some few remains of the late Norman church surviving. (p. 107.)

Wingfield, South (1571), contains the old Manor House, an interesting ruin known as Wingfield Manor, built in Henry VI's time, where Mary Queen of Scots was imprisoned. (pp. 7, 86, 87, 122, 132.)

Wirksworth (3807), 13 miles north-west of Derby, is situate in a valley at the southern extremity of the lead-mining district. An ancient brass dish is kept at the Moot Hall as a standard for dishes for measuring lead, and the barmote courts for swearing in the Grand Jury and settling mining disputes are held here. (pp. 12, 17, 48, 76, 78, 80, 83, 89, 95, 103, 106, 107, 111, 131.)

Youlgreave (1077) Five miles from Bakewell. The church of All Saints is of mixed styles and includes some Norman work. (pp. 79, 102, 107, 109, 126.)

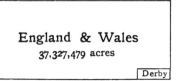

Fig. 1. The Area of Derbyshire (658,885 acres) compared
with that of England and Wales

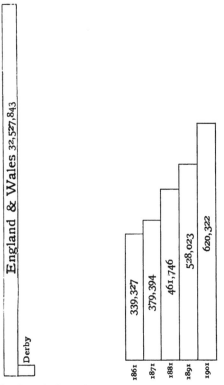

Fig. 2. The Population of
Derbyshire (620,322) compared
with that of England and
Wales, 1901

Fig. 3. Increase in the
Population of Derbyshire from
1861 to 1901

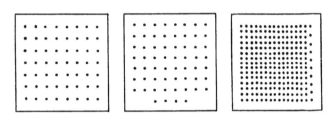

England and Wales 558   Derbyshire 600   Lancashire 2347

Fig. 4. Average Population to the sq. m. in England and Wales, in Derbyshire, and in Lancashire in 1901

(*Each dot represents ten persons*)

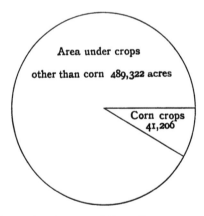

Fig. 5. Proportionate area of Corn Crops to area of Crops other than Corn in Derbyshire in 1906

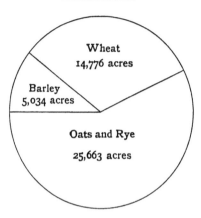

Fig. 6. Proportionate area of Wheat, Barley, Oats, and
Rye in Derbyshire in 1906

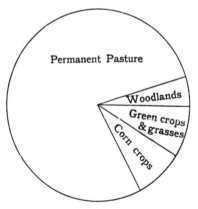

Fig. 7. Proportion of Permanent Pasture to other
Areas in Derbyshire in 1906

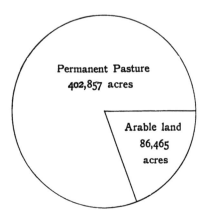

Fig. 8. Proportion of Permanent Pasture to Arable Land
in Derbyshire in 1906

Milton Keynes UK
Ingram Content Group UK Ltd.
UKHW032321161024
449665UK00001B/11

9 781107 637337